The Way of Love

on the Camino de Santiago

By

Angela Leslee

Cover design by: Douglas Webster

Poppy art on back cover by: Cooper LaMere

Published in the USA by: Angela Leslee

ISBN: 978-1-79325-299-9

Dedication

For my beautiful grandchildren.

You all hold a special place in my heart.

May you embrace adventure as a way of life.

Table of Contents

Relationships - The Final Frontier

Some might say I've been unlucky in love. Sure, I've been divorced twice, who hasn't? But then three of my next four relationships all ended eerily in the same manner. All declared their love for me, but for various reasons, they couldn't commit and had to move on. Like my own personal *Groundhog Day* movie, each one echoing the next, "It's not you, it's me." Not very original, but noble, I'll give them that.

The real kick in the butt however, the part I couldn't quite let go of, each one then went on, in the year following, to marry the next woman they met. True story, I couldn't begin to make this stuff up. The last was 13 years ago. I really thought he was *The One*, we had so much in common. But turns out, he was someone else's *One*.

I may be a slow learner, but I eventually figured out that of course, it was me and not them who was afraid of commitment. But by then it was too late, the damage was done. I decided *enough* with relationships and all the time and energy they can suck. It was time to get a dog and get on with my life.

But that isn't why I decided to walk 500 miles across the entire country of Spain by myself – at least, I didn't think that was why.

And so, it begins...

Douglas retrieves my backpack from his trunk, then turns to give me an all-encompassing bear hug complete with a kiss, square on the lips. Fifteen years ago, we ended a two-year relationship, now we are technically just friends, no benefits. Most of the time our kisses are platonic and land on the cheek. But when the mood arises, in tacit agreement we kiss on the lips, acknowledging the ghost of intimacy that remains between us. He was the only one of the four that remained stoically single.

Douglas has accompanied me on many of my training hikes in the forest and along the beach. It gave us time to re-visit the love and tenderness of a soul connection outside the boundaries of a relationship. He surprised me with his knowledge of athletic training and gave me several good endurance tips. His once slim body, now hidden under extra pounds, caused me to forget he was an Olympic class swimmer in college.

With a wistful smile he says, "I'll be following your blog, happy trails," and pulls away from the curb. I watch for a moment and wish briefly he had taken me up on my offer to accompany me. I turn, heft my pack over my shoulder and whisper to myself, *and so it begins*.

It turns out that traveling with, and taking an 18 lb backpack as carry-on luggage, is an experience unto itself. I'm fairly well traveled, so I have my own system for going through airports. But I am quite unprepared to handle this monkey on my back. Like a *Three Stooges* act, the man standing next to me at check-in has to duck and move quickly when I turn without warning.

As someone who compulsively worries about what people think, I can't help but wonder at the sight of a 62-year old woman sporting big hiking boots and a large, overstuffed backpack. My shoulder-length white hair makes me easy to spot in a crowd. I don't *blend*, which makes me feel all the more conspicuous.

As I approach the conveyor belt in the airport security line, I snap out of it and realize I have to get myself out of this contraption. I unsnap the buckles and let the pack roll off my left shoulder. Oh crap, where did I put that baggie with the liquids? I thought I placed it on top for easy access. Digging thru my pack only makes matters worse, I start to panic. I lock eyes with the stone-faced TSA lady, apologize and smile weakly. She raises her eyebrows and crosses her arms over her chest, letting me know she has all day. Nothing for it but to empty everything out on the conveyor belt. With the entire contents strewn about for all the world to see, I take a quick look behind me. My fellow travelers are clearly not amused. I feel the sweat start to bead on my forehead. Then I remember, I put it in the small compartment at the top so that this wouldn't happen. As I shove everything back in, my once neatly packed bag is now a jumbled mess. At this point, I'm going for expedience over organization. With a bored look, the TSA agent nods at my tightly laced boots. I follow her gaze, "Oh," I say with a sigh, "these have to come off too, don't they?"

As I lace my boots back up, I admire the angels that Jean Luc, an artist friend of mine, painted on the toes. Every time I glance down at them, they are to remind me that my angels will be looking over me to help me safely traverse the Camino de Santiago.

The Camino de Santiago is a network of pilgrimage routes in Europe, laid claim to by the Christians 12 centuries ago. All roads, like spokes of a wheel, lead to Santiago de Compostela in northwestern Spain, where legend has it the remains of St James were discovered.

In ancient times, a pilgrimage started at your front door. Some Europeans continue this tradition by walking from their homes across Europe. Today, however, the most frequently traveled pilgrimage route is the Camino Frances, with more than 200,000 pilgrims per year completing it. The traditional starting place for the Camino Frances is in St.-Jean-Pied-de-Port, a medieval village in the foothills of the Pyrenees, in south-western France. It quickly crosses over into Spain, then continues on to the ocean on the other side of the country. The route is 800 kilometers, roughly 500 miles, and aligns with the Milky Way.

While considered a Christian Pilgrimage, it is thought to pre-date Christianity. So, it's difficult to know for sure how long this route has been traveled and how many souls have walked in these hallowed steps. That it is steeped in significant, spiritual energy goes uncontested by most. It has been traversed by paupers and Kings, by Holy men and laymen. It was once guarded by the Knights Templar. Some of the original Roman roads remain.

This journey is undertaken for countless reasons, but there is often a religious or spiritual, underlying purpose. Due to its longevity and popularity, the infrastructure is well established. A thriving cottage industry has sprung up in the many small, ancient villages en-route providing shelter and food for pilgrims. I'm not exactly sure why it is *I'm* doing this, only that I have a strong desire coupled with my sense of adventure.

I walk through into the open-air waiting area, one of the things I love most about the airport in Kona, Hawaii. The warmth of the sun on my face restores my sanity. I take a deep breath and a sigh of relief. After 22 years of living here, this airport is as familiar to me as my kitchen. Why then today, do I feel awkward and unsure of myself? What happened to all that self-confidence I can usually fall back on?

Oh God, I don't have time for self-reflection right now, this journey has barely begun, and I need to pee. In the lady's room, I head for the nearest stall. I chuckle to myself over the little drama that TSA angst brought on. As I turn to get around the door, I am once again stopped short by the bag strapped to my back. Now, I laugh out loud, like some crazed homeless woman, which, as I catch my reflection in the mirror, I bear some resemblance to. I head for the wide door of the handicapped toilet and try to sit down before I wet myself laughing.

Bonjour

Flying halfway around the world is a grueling, process. At the last minute, I decide to upgrade the nine-hour, Chicago-to-Paris leg to Business class. Best move ever!

Thank goodness airline attire is more casual these days. My backpack and hiking boots are less conspicuous than they would have been 30 years ago. I try to not look star-struck as I gaze down at the individual pod with my number above it, complete with side-table and a seat that reclines completely flat. I've seen these only in movies. My I-do-this-all-the-time facade is busted when I can't figure out how the seat works and must ask the flight attendant for help.

I have not slept in 22 hours, so I opt for the *quick meal*, instead of dragging it out for an hour to duplicate a fine-dining experience. I push the remains of dinner aside, recline my seat all the way back and proceed to sleep like a dog on valium. I don't come back to consciousness until I feel the flight attendant shaking me lightly, offering me a warm washcloth with tongs as we descend into Paris.

I have nine hours until my flight to Biarritz. I was so happy when I booked this that I would not have the added stress of feeling rushed. After a short walk, I reach my gate. I find it odd I haven't gone through customs yet. But it's been a long time since I traveled internationally. Maybe that doesn't happen until the next airport. I shrug my shoulders and settle in for the wait.

The airport is quiet at this early morning hour. I relax at a coffee shop, order a café au lait and a pastry, and get comfortable with my phone to read an e-book. My iPhone will be my entire entertainment and communication tool during my seven-week trip.

I'm well into my book, and the time has flown, but I grow bored. A little birdie tells me to check on the status of my flight. Hmm, weird, it's not listed on the departing flights, and it's only a couple of hours until take off. After more investigation, I see that I have to go down a narrow hallway to get to my gate. Voila! There is Customs, with a long line snaking around. Deep breath. After all the waiting, I now have a sense of urgency. Considering the current political climate, I don't dare allow my panicking, *crazy lady* to have a voice. What would she say anyway? "Excuse me, can I please cut to the front of the line? You see I screwed up, and I've been sitting for seven hours reading a book instead of making sure I'm in the right place." – No, I don't think that would fly. So outwardly calm, I await my turn.

As I tuck my passport safely back in my bag, I review my next options. A small corridor to the left looks like an employee exit. The large pair of heavy-duty, closed doors in front of me seems the more obvious choice. However, they look like one-way doors, and there will be no return through them if I am wrong. My confidence has been shaken from the recent mistake made by a travel-weary mind. I'm frozen in indecision. There is an officer armed with a big rifle, standing solemnly nearby. I don't know if he speaks English, so I decide to keep it brief. "Gate 23?" I say as I point towards the doors.

"Bonjour," he says reproachfully, with a slow nod of his head and waits. My face reddens. I am being chastised for acting like the rude American. This learning will serve me well over the next seven weeks. I will discover that Europeans are very polite to each other, and all interactions, no matter how brief, start with a greeting.

I try again, "Bonjour and pardon." He nods like he is the Buddha imparting some great life lesson on me, and he points towards the door.

Holy cow! Once inside the door, a noisy crowd swallows me up, and I have to work hard to stay in my body with all the busyness and mayhem around me. After a ridiculously, long walk, I come across another airport security checkpoint. I wait in line again. Following that is another hike, that seems to take forever. Finally, I am at my gate and people are already starting to board.

With a big sigh of relief, I take my seat on the plane and look out the window to watch them load luggage below me. I reflect on my most dazzling blonde moment to-date, the narrowly averted disaster of potentially missing a flight with a nine-hour layover. I roll my eyes, shake my head and snort a little laugh.

After a short flight, we land in Biarritz. Out the window I see several military-looking officials with rifles over their shoulders, pacing on the blacktop. Hmm, interesting. Everything feels so foreign. The adventure part of this is exciting, but I also realize how very alone I am.

In the small airport that is blessedly quiet, and free of confusion, I feel I can finally breathe. I see other pilgrims for the first time. They are easy to spot with their shiny, new backpacks. I am finally here and can now consider myself an official pilgrim. I will come to find out, that most of the Spanish people, especially those who live along the Camino, are very respectful and helpful to anyone choosing to walk their time-honored spiritual pilgrimage.

At first, we keep to ourselves. We are not yet in the Camino mode of striking up conversations with complete strangers. But resistance is futile. Our unacknowledged common bond draws us magnetically towards each other, and we begin tentative conversations, mostly concerning transportation to St-Jean-Pied-de-Port.

Eight of us share a taxi for the hour-long ride to St. Jean. Our driver is a local beauty, with long black hair and a mini skirt. She loves to practice her English by proudly telling us stories about the Basque countryside we are winding our way through. I notice, in my out-loud voice, that all the houses have identical color schemes; white with red roofs and trim. She says, "This is how it's done in Basque country," - as if that is explanation enough. To this day, it is hotly debated that the Basques are neither French nor Spanish, but their own country with their own language and culture. I will come to see that they suffer the same plight as Native Hawaiians, fighting for their sovereignty.

As we approach our destination, our driver asks us individually where we are staying. Why she asks, I don't know. Because, after an hour of driving, she pulls up on the side of the road in St-Jean-Pied-de-Port, in front of nothing in particular, and unloads our luggage.

With the flourish of a flight attendant doing the *exits* drill, she indicates that we should all disembark.

I'm afraid I have missed something, so I ask her where I am to go. She waves her hand vaguely and rattles off something unintelligible in French. Her mission complete, she is ready to be on her way. With an "au revoir" and a big smile, she drives off. My fellow cab-mates and I look at each other, a little dazed and confused. My 32-hour travel day is not over yet.

The delicious beauty of this medieval town that sits at the foot of the Pyrenees in France is not lost on me, despite an exhaustion that's threatening to consume me. I follow several other pilgrims, hoping they know something I don't. The narrow cobblestone road doubles as a walking path, so we have to move aside as a car slowly drives towards us.

As we approach the Rue de Citadel, I admire a large, stone archway over the road to the right, and beyond it a one-lane bridge. Picturesque does not begin to describe this town straight from the pages of a fairy tale. I would love to explore but need to find my albergue for the night. Similar in concept to hostels, a room in an albergue may contain anywhere from two to thirty beds. However, along the Camino albergues are reserved for pilgrims. I turn to the left, still following the others and am confronted with another cobblestone road, but this one is crazy steep. Well-cared-for brick buildings that are ancient as dirt, four stories high and all attached to each other, line the street and open their front doors right onto it.

My mind relaxes, and my body feels alive as it sinks in that my adventure has begun. It does feel a little strange though, that in such a short time I've already had so many experiences and no one to share them with yet. But perhaps the containment of this adventure adds to its' potency. Reminiscent of zealously guarding those first precious hours after giving birth, when you don't want anyone to dilute the powerful *first kiss* you have with your new baby. The multi-layered, multi-cultural richness of this experience is becoming difficult to put into words.

A travel zombie, deep in thought, I walk half-way up the street before I realize the foreignness of everything has distracted me from looking for my albergue. Pause – rewind, I turn to go back down again. Thankfully, after only 20 feet, I notice a tiny, post-card sized, hand-painted sign. It announces that this unassuming four-story, centuries-old house will be my home for the first two nights.

A narrow, open door beckons me into a long entryway with hooks on the wall above long benches. At the far end of the foyer, a sign tacked onto the door leading into the house, proclaims it is locked until 1900. I gladly relieve myself of my backpack and hang it on one of the hooks, as I see others have done. A hip bag that never leaves my sight, containing all my valuable documents, phone, and money, remains securely fastened around my waist.

32 hours of travel has left me in a weird, limbo-like state. I'm not really hungry, but I decide to eat anyway, hoping it will ground me. France is only on the itinerary for two days, I would love to sample some famous French cuisine. But that is not in the cards for this tired pilgrim tonight.

In preparation for this trip, I have been brushing up on my high-school Spanish. But upon my arrival in France, it became immediately apparent that I only know about three words in French. A menu posted on the wall of one appealing café has no English translation and is unintelligible to me. At another restaurant, there is a menu scrawled in English on a blackboard placed at the curb. This one is a beehive of activity, with pilgrims everywhere, laughing and conversing in many languages as they enjoy their meals. But I am prepared to pay for a nice dinner tonight. There will be ample opportunity to partake of the ubiquitous *Pilgrim Meal*, which this restaurant is offering. I have read that this budget-conscious, three-course, heavy-on-the-carbs meal, will be offered everywhere on the Camino, so I continue on.

At a sweet café with an awning, a waiter in black pants and white shirt sets up empty tables outside. I follow him inside as he goes behind the bar. "Bonjour! Do you have a menu in English?" I ask politely. He dismisses me with a disdainful look and a curt, "No." You would think I asked if they served hot dogs.

Hmm, I have heard the French aren't particularly fond of Americans. This must be what it looks like. Across the street is another quaint café, and following my repeat query for an English menu, this waiter responds in a similar manner. He curls his lip as if I've asked him to take off his clothes and run naked through the restaurant.

Suddenly I remember my Google translator app. But as I rummage in my hip bag for my phone, I see how this will go down: *Excuse me while my app loads. Oh, and by-the-way do you have WiFi? And what's the password?* – I watch his retreating back as he leaves to serve other customers. Those polite enough to learn at least enough French to read a menu before traveling to his country.

Well, that was enough rejection for one day. I guess there's no time like the present to embrace my pilgrim status. The busy bar I had turned my nose up at earlier is sounding much better. As I sit a little awkwardly at an empty table sipping a glass of red wine, a woman who was in my taxi from the airport, approaches me with a big smile. We greet with a hug like old friends, although we didn't say two words to each other on the way to St. Jean. Her name is Clare, and she joins me for dinner and some wine. She brushes her long, blond hair out of her beautiful face as she lights a cigarette. Her British sense of humor and deep belly laugh is exactly what I need to unwind, on this, the longest of days. I forgive her for blowing smoke in my face while I eat.

I return to the albergue. The inside door is now invitingly open, and my backpack is hanging all by itself. The young, pretty French woman that runs this lovely private home knows no English but greets me with a smile. My tour consists of finger pointing and French words I have to guess at. She shows me the kitchen where breakfast will be served. We then walk up three flights of stairs, where she points at an open door then turns to leave. The small attic bedroom with a steeply sloped, head-smacking, ceiling has four beds. Another woman and two men barely acknowledge me as they continue preparing to retire for the evening. I know from my reading that communal living arrangements are the rule rather than the exception on the Camino, but knowledge and reality are two different animals. Fortunately, I'm too tired to process this now.

With a travel-weary mind, I try to look like I know what I'm doing, as I go through what will become a daily ritual of unrolling my sleeping bag and preparing for the night. I'm glad I followed the suggestions of other pilgrims and decided to sleep in my clothes (to save weight in my pack) as there is nowhere to change. It is starting to sink in that for the next forty days nothing in my life will feel familiar. As I drift off to sleep, I notice the little girl inside me sucking her thumb and asking, "is it time to go home yet?"

Day 1: Orisson

Someone else's alarm goes off, but once again, I am already awake. Yesterday on my day off, jet lag mandated a 3:30 a.m. wake up. At 5 a.m., the man sleeping opposite me had rolled over to silence his alarm. Then all three of my roommates had wordlessly rustled around, packing for an early departure. A day off to recuperate from travel sounded like a good idea when I booked my trip. But instead, I felt like the kid who's kept behind during recess because she hasn't finished her assignment.

But today it's my turn.

Shortly on the heels of deciding I needed to walk the Camino, came the realization that I would be walking alone across an entire country. The journey would take me over mountains, through forests and miles of open countryside. I would be armed only with basic knowledge of the language, and with no car to jump in and lock the doors if things got sketchy – gulp! I have always been cautious to not put myself in dangerous situations. This could potentially be the Grand Mac Daddy of dangerous situations! The patronizing, prevailing wisdom on the Facebook forums I frequented daily, was: *you're never alone on the Camino,* because so many are walking it these days. But I knew that no one could guarantee that, and indeed later found that was correct. You actually can be completely by yourself, sometimes for hours, in the middle of nowhere.

Having a partner who enjoyed participating in adventures with me, would come in pretty handy right about now. But since that train has left the station, I began begging, pleading and cajoling my friends and family to join me. "500 miles? whaaaat?" "Wow, that's a cool idea, good luck with that." I was usually dismissed with a look that said I'd taken leave of my senses. Intuitively, I knew I would have the most potent experience if I walked by myself, but I was just plain scared to.

Three months before I was due to leave, I saw a letter on a Facebook page from Mary Jo in Tennessee, asking for female partners to join her walking the Camino. Her dates were perfectly aligned with what I had envisioned, so I contacted her. She already had three other women in her posse but welcomed me to join the group.

We were from all over the United States, but through the magic of Facebook messenger, were able to meet and get to know each other's quirky personalities. I now had four other cohorts that didn't think it was strange that, in the interest of an uber-light backpack, I weighed my underwear.

Amidst the relief of feeling some security that I would not be alone, however, was a niggling doubt in the back of my mind. I knew human nature too well and reality rules. From all my reading, I gleaned that the essence of the Camino is very fluid, with lots of coming and going of people and relationships. How could I be sure that these women would actually walk the entire way with me? I pictured my angels smiling down on me, "whatever it takes to get you there and walking," they had said with a wink.

So, this morning I look around at my fellow *Hags with Bags,* a moniker for our group irreverently suggested by Deidre's son. We all met yesterday afternoon after I spent a relaxing day exploring St.-Jean-Pied-de-Port, a town so sweet it makes my teeth hurt a little.

Mary Jo is first to get out of bed already wearing her ear-to-ear smile. She runs her fingers through curly, brunette hair, peppered with grey, tying it in a ponytail. Although she is my age, she is tiny, and her petite figure gives her the illusion of perpetual youth.

"I miss my dogs," are the first words out of her mouth, followed quickly by "and my husband haha."

Deirdre, talking auctioneer fast, even at this early hour, pipes in, "I can't think about my husband and kids too much, or I'll have to go home." Then she too laughs, a quiet, staccato *rat, tat, tat.* Deirdre's tall, thin frame matches her face, which can go from worried to exuberant in a blink. She's a tad high strung, but I love her enthusiasm.

Camilla's the quiet one of our group. Her beautiful, Asian features wear an unreadable mask as she gets up to use the bathroom. At the last minute she decided to bring along a friend, Jean, who's staying in another room. Jean exited the taxi yesterday looking scared and unsure, with a force field around herself saying, *stay away*. Last night Camilla had given us Jean's story, since she had sat quietly at dinner not speaking a word. Jean is care-taking her ailing husband, so life at home is difficult. Camilla thought the Camino would be good for her. I think she was also supposed to be Camilla's back up plan in case our group didn't work out. But Jean ends up separating from us all after the first day – the shit hits the fan really fast on the Camino. I will meet up with Jean in four weeks though, and she is transformed. Turns out Camilla was right, the Camino *was* good for her. She shed 20 lbs, found her own *Camino Family* complete with *Camino Husband* – but more on that later.

Before dinner last night, we made a group trip to the Pilgrims Office, where we signed in and got our credencials (pilgrim passport). The accordion style document will collect sellos (stamps) all along the way from albergues, bars, restaurants and churches, as proof that we completed the journey. This then allows us to get our Compostela (certificate of completion) in Santiago. At least one sello per day is required, although many days I will get more. We are all now official Pilgrims.

The fresh French bread for breakfast with coffee is delicious. The crustiness of real French bread isn't so hard you're afraid it will make your gums bleed. Just a delicate crunch, then soft, yeasty amazingness. But with the change in schedule and food over the past 2 days, my *system* is out of its usual rhythm. According to all sources, we will be hiking for several hours this morning with no access to a bathroom. Unfortunately, my hands and apparently my intestines are tied, there's nothing to be done about it. I put it out of my mind.

What's not to love about bright blue skies with cotton candy clouds. It's chilly though, the temperature is 50°F. We know this first day is a steep climb, straight up into the Pyrenees mountain range, so we dress for a Nor'easter. I put on a short sleeve T-shirt, long sleeve

T-shirt, fleece jacket and light rain jacket, everything I brought with me except my poncho.

Our pre-arranged meeting with Pamela, the last of the *Hags* in our group, goes smoothly. She has pre-booked all her lodging, knowing that we plan to wing it. I guess correctly, that we won't be walking with her for long. But for now, the camaraderie of our little gang helps quell the nervous jitters at the start of such a big adventure.

On the outskirts of St. Jean is a large, colorful sign announcing the beginning of our journey. I discovered this yesterday when I had another run-in with Clare all geared up, ready to leave. I gave my new best friend a big hug and walked her to this spot. Momentarily, I wished I could have spontaneously taken off with her. I got a foreboding of the price I may pay for trying to control this trip.

"Buen Camino!" I had said for the first time, as I bid Clare farewell. The words sounded strange rolling off my tongue. I'm really here. I'm really doing this. The universal greeting/blessing *Buen Camino,* given by one and all to pilgrims on the Camino, will soon come as naturally as a *bless you,* to someone sneezing.

After the half-kilometer walk to the sign, we are all sweating already. So, we take off everything down to our T-shirts. I tie my long-sleeve, hooded T-shirt around my waist for quick access, in case I've misjudged. After re-situating our packs, we rope in a pilgrim walking behind us to take a group shot.

For 25 years, I have held the belief that healthy bowels lead to good health. My daughters call it; *obsessed with poop.* I have done every colon cleanse in the book; consequently, the good news is I'm very regular. The bad news is, when I have to go it's non-negotiable. One of my concerns with walking the Camino was the miles and miles between bathrooms each day. After much deliberation, I came armed with doggie poop bags and a plastic backpackers shovel – I like options. Although prepared, I am concerned with the real-life logistics of this maneuver. I pray I can get used to the myriad of new things going on before I have to deal with it.

But no! Thirty minutes into the first day, my stomach starts its familiar rumble. Un-freaking-believable! I guess the walking got things moving again. I look ahead at the endless country road, with farmer's fields on either side, and the occasional house dotted about, with no large tree to hide behind. Seriously? This was *not* part of the equation.

I don't want an audience; fortunately, Deirdre and Pamela are fast walkers and well in front of the pack. Camilla and Jean lag behind Mary Jo and me. Maybe I can ignore the urge and it will go away. "Umm, nice try," says my tummy, "gurgle, gurgle." The rumble is more than an urge, it's a pain that doubles me over and threatens to make a mess of things if I don't pay attention.

Reality rules. Sigh! What comes next is evident, as I roll my eyes and begin to sweat. There's nothing for it but to confess to Mary Jo about my situation to enlist her help. We look around for a somewhat-private place, as Camilla and Jean catch up. Great! More people to witness my shame. But there is no private place. The four of us continue to walk. It becomes clear I'm running out of options. Then Mary Jo spies a country dirt lane off to the right, with the grass flanking it about eighteen inches tall. It's not ideal, but it will have to do.

As I move fast away from the main road, I realize I haven't even had the chance to master *peeing* with an 18 lb backpack throwing me off balance, let alone this. I will later learn to remove my pack before trying any squatting maneuver, but it's all I can do to make it to a somewhat private place in time. Forget digging a hole in the hard clay earth, the shovel looked good on paper but will be ditched in a few weeks, unused.

I teeter on the balls of my feet and try not to topple over. My relief is short-lived however, as I glance down and notice the arms of my long-sleeved, hooded T-shirt tied around my waist. *OH MY GOD, the hood!* I envision it hanging behind me, between my legs, creating a perfect basket. The possible ramifications are unthinkable. What's worse, I stand and pull my pants up, but can't see around my backpack to assess the potential damage to my only long-sleeved shirt. I am afraid to reach back and touch it, just in case – well, you know, just in case…

I rejoin my friends, and with bated breath, I explain my dilemma to Mary Jo and ask her to check my hoodie. With a little laugh she says, "it's fine." Whew! Now I can laugh too. And laugh we do, until we cry.

The road leading steadily up is a gentle climb through the gorgeous Basque countryside. The first couple of miles, fueled by joy and enthusiasm at finally beginning our epic journey, is so much easier than I had imagined. As we get higher, the views have an otherworldly quality. The Pyrenees mountain range, for as far as the eye can see, blooms green with the advent of spring.

Deidre and Pamela are now out of sight. Camilla hangs back with Jean, trying to engage her, but it's obvious she wants to be left alone. Mary Jo and I share the same pace, the same fear of walking solo, and a similar sense of humor. We agree to stick together as the group dynamic shifts and changes.

Three kilometers before our destination, the ascent steepens. My legs burn, and my body feels heavier as I push off with each step. When I stop to take a breath, I have to brace myself on my poles so as not to topple backward. The panorama begs picture taking, affording us the opportunity for many breaks.

As we climb higher, we stop at a spectacular vista. I make sure that we are alone on the trail, then ask Mary Jo to video me. I look into the camera and say, "This one's for you Douglas." I look out across the valley to the opposite mountain and yell, "THANK YOU." Whenever Douglas's cup runneth over, he will stop wherever he is and do this, no matter who is around. Often this is in the middle of the ocean while we are swimming with dolphins. In spite of it being a charming quirk, it often makes me cringe and look to see who's watching. I really need to get over worrying about what other people think.

Can this climb get any steeper? My legs are burning. Mary Jo spots an old, sprawling tree at the side of the road. She clambers down and sits against the ancient trunk for a photo and discovers a tiny statue of Mary and baby. Excited by the discovery, I climb down to take my own picture, and to say a prayer for my grandson Cooper.

Two years ago, when he was four-years-old, Cooper was diagnosed with leukemia. It has been a heart-wrenching time for the whole family. Even though he is in remission and expected to make a full recovery, the three-year treatment plan of daily chemo is brutal. When I decided to do this walk, I recognized it as an opportunity to use the spiritual nature of the Camino to say a daily prayer in power spots for him.

Soon after we are rounding a corner, and like a mirage there stands our destination, looking uncannily like a ski lodge. It is still early, before noon, but the others in our group are waiting for us. We check in and spend the rest of the sublime afternoon lounging on a deck overlooking a picture book setting. The faint music of cowbells in the distance serenades us as we enjoy wine and congratulate ourselves on our successful first day.

Dinner is a communal affair at Orrison. Trestle tables are set up in three rows to seat the 50 pilgrims lucky enough to score a bed. Most of us booked months in advance. The alternative is to make the long, arduous hike over the mountain in one day, which of course, many do. I am glad to have a gentler start to the trip. During dinner we are asked to take turns standing, to introduce ourselves and explain why we are making the journey. The ritual is meant as a way for people to meet, and encourage a sense of community, of being part of something bigger than ourselves.

Day 2: Roncesvalles

And then there was snow. And hail. And sleet. And rain. And 50 mph winds! What a difference a day makes. Although we leave as a group, Deidre and Pamela strike out in front and are soon lost to us. We plan to meet them in Roncesvalles. Mary Jo and I look at each other as we begin our hike up the mountain in the dreary weather. We are as naïve as brides on their wedding day.

"I've been thinking about this for so long," I giggle.

"I know. I can't believe we're finally here." Her broad smile tells me she's as excited as I am.

It feels like I'm walking in honey, on legs stiff from yesterday's climb. As we get going, the weather grows worse. The wind gets stronger, and the rain beating against our ponchos turns to hail. My leg muscles eventually warm up, and in spite of the cold, it becomes easier.

The relentless ascent continues. As we turn a corner, a gale-force blast of wind takes my breath away. The sun makes a brief appearance to reveal that this disembodied path in the clouds we have been following, sits high above a remote valley. Then miracle of miracles, a rainbow! I remove a glove with my teeth and struggle with my poncho to get to my hip bag, buried under two layers of clothes. Before I can get to my phone, the clouds reclaim the rainbow. I will have to keep that picture in my head.

The wind picks up even more and whips our ponchos around us. The vinyl flaps against my face and ears making communication difficult. With wide eyes we laugh and plant our poles and brace ourselves against another huge gust.

The hail morphs to sleet and then to snow, which is beginning to accumulate. The wind threatens to knock us off the path. This is suddenly no joke. My naïve bride wakes up in her suburban house, three years and two kids later and cannot find her rose-colored glasses. I can see why the walk over the Pyrenees, is touted by many to be one of the more difficult days on the Camino. Tomorrow, this pass will be closed to pilgrims, due to the weather. Today, ignorance is bliss.

Mary Jo yells to be heard over the wind. She points out the Virgen de Biakorri, a famous statue of Mary and Child. The knee-high figurine sits atop a rocky throne. We step off the trail to admire her and are joined by Camilla and her new walking buddy Marie. Jean appears to have vanished into thin air. As we stand in front of the iconic sculpture, the snow stops and the clouds part, to reveal a startlingly blue sky with brilliant sunshine. I seize the moment for a quick prayer for Cooper. We take our pictures and turn to leave. The sky which had miraculously parted, clouds over and a slurry of snow and hail sting my face.

"I think we just witnessed our first Camino miracle," I say to Mary Jo.

The wind is so strong, it sometimes feels like we might lose our balance and be blown off the mountain. I turn to Mary Jo and smile "I feel like we're getting the full Camino experience today." She laughs in agreement.

We enter a magical forest. Trees naked for the winter, nestled in their own rotting leaves, send up a faint whiff of earthy decay. They flank either side of a path on even ground. The wind has calmed down, a dusting of snow coats everything. The silence, but for the crunch of our boots, adds to the ethereal surroundings. A welcome respite, it feels like walking in a park. A sense of exhilaration leads me to believe my endorphins have kicked in. The final push uphill to the highest point of our 2,000-foot ascent where a foot of snow has accumulated, brings Mary Jo's inner child out to play, she lays in the snow and makes a snow angel.

Ahead of us is a tall, overweight man who struggles with his unruly poncho. The strong wind has returned and broken one of the snaps, he tries but fails to anchor it. As we get closer, he looks at us. Wordlessly his wide eyes say: *Help! My poncho is out of control, and I can't remember why I'm doing this.*

I recognize this middle-aged man as David, we met him at dinner last night. He shared how his wife gave him a send-off party and fully supported his endeavor but respectfully declined to join him. Everyone had laughed. After helping secure his poncho, David joins us. His pace matches ours, and he appears to be in need of a little company.

Snow covers the landmarks and creates some confusion, as we come to the part in the trail to begin our descent into Roncesvalles, where a large monastery will be our home for the night. There are two ways to go, one beautiful but slippery, steep path through the forest for two kilometers. The other route is a switchback paved road, half a kilometer longer, still steep, but much safer.

After some deliberation, we end up on the road we want, the longer, safer one, and begin the precipitous descent. I feel a blister forming on my baby toe, but it cannot take away from my excitement as we spot the monastery from on top of the mountain. It beckons like a fairytale castle nestled deep in the forest below.

We reunite with Deidre and Camilla at the monastery. After a 25-minute wait, a hospitalero assigns bunkbeds in the newly renovated, 180-bed albergue. A top bunk is the last to be claimed, as I enter our four-bed cubicle. Only a couple of rungs on a vertical ladder separate me from my bed. But it's surprisingly difficult to heave myself up there. This will be no fun in the middle of the night when I need to use the bathroom. Deidre, younger and more agile than I, watches my heroic effort to scale the ladder and immediately offers to switch bunks. Bless her heart.

As advised, we check for bed bugs, but what would we do if we found them? There is nowhere else to sleep. In fact, everything, including the small hotel Pamela reserved, is *completo* (full) by 2 pm. What will the pilgrims behind us do for the night?

I have brought so few things, all neatly folded and separated in gallon zip lock bags. I was under the smug and false impression this would organize my backpack despite the few compartments. But deja vu. Like last night, I empty the whole damn thing on my bed to find my elusive toiletry bag and clean clothes so that I can shower. This is the first and last albergue I stay at, with segregated showers for men and women.

After my shower, I attempt to organize my backpack again. For God's sake, I carry clothing for two days, rain gear, and the barest essentials of toiletries. How hard can this be? Turns out, I will repeat this futile exercise daily for the remainder of my trip.

It is a novelty for me to eat chicken, as I have been a vegetarian with few exceptions for 25 years. However, to maximize my personal growth potential, I decided to drop as many habits, routines, and beliefs as possible on this journey. Throw caution to the wind and see what happens! Vegetarianism fell into all three categories. In one way or another, food has served as a primary focus for most of my life. From dieting to vegetarianism to veganism to cleansing to extreme healthy eating, what I eat has defined me. I decided long before I left home that I did not want this trip to be all about food. In my preparation for this journey, I learned about pilgrim meals and menus that would be available. It is easy and economical to eat along *The Way* (a common Spanish colloquialism for the Camino) if you go with the program. So, I decided I would eat what was offered to me with gratitude – within reason.

I forgot how good chicken tastes! No wonder it is the yardstick by which all meats are measured. *Tastes like chicken*, the universal complement and green-light to try something different.

Between the long day and late dinner, traditionally eaten after 7:30 p.m. in Spain, we are more than happy to retire early. Our sleep cubicle offers a modicum of privacy, more than will be available at most other albergues. But as I drift into a deep, dreamless slumber, I hear a surround-sound chorus of soft snoring.

Day 3: Zubiri

In retrospect, we should not have waited for breakfast. It set us back at least an hour. But we had already paid the night before. God forbid we should donate those three euros to the monastery and be spontaneous this morning.

We turn the corner at 7:30 a.m. to the small bar meant to serve 200 people and see the line out the door. A light dusting of snow fell overnight, pilgrims stamp their feet while waiting in the cold. Twenty-five minutes later, we are inside the door. The line snakes around the tiny room, but we are glad to be warm. I see what the holdup is. One bartender furiously works the handles of an espresso machine. Is she making an individual espresso and steaming the milk for everyone's café con leche? The Spanish take their coffee even more seriously than we do in Kona.

I am standing right next to the dining room door as it opens and a group of people leaves. My curiosity and impatience get the better of me. I step towards the door and open it a crack to peek inside. Without missing a beat, the barista leans across the bar, eyes blazing, scolding me loudly in Spanish, while pouring steaming milk into a cup. A handsome, young Irishman in front of me turns and says with a cheeky grin: "Well, that's you sorted then." I'm grateful to him for diffusing an embarrassing situation with laughter, we both crack up.

After an hour in line, the fresh squeezed orange juice, hot, *fresh* café con leche and a large slice of toast and jam taste good. But it was hardly worth the wait. As we gear up to leave, I belatedly wonder if we should have skipped breakfast and just started walking this morning.

Pamela left thirty minutes ago, she didn't have to wait so long for her breakfast in the Hotel. Our group of four (Jean is gone, but Marie has now joined us) spreads out naturally. Mary Jo and I have a good partnership and travel together again today. We begin with a lovely walk, in woods famous for witch covens in the 17th century. There is a white cross monument with a placard telling us it commemorates witches burned at the stake and pilgrims. I'm unsure of the relationship between witches and pilgrims. However, after reading *Hidden Camino* by Louise Sommer, I've become very curious about the pre-Christian origins of the Camino and their possible connection to Goddess worship societies. Leave it to me to be interested in the part no one's talking about.

After several gentle kilometers (read: no hills), Mary Jo and I come upon the quaint village of Burguete. We stop for second breakfast, as we have quickly burned off the meager slice of toast. Second breakfast is a common practice on the Camino. My meals on many days will consist of two breakfasts and dinner.

As we leave the bar and walk down the cobblestone street, a yellow arrow on the road guides us to turn right in between houses. In 1984, the brainchild of the yellow arrow system, Don Elias Valina, a parish priest at O'Cebriero, drove the Camino Frances painting yellow arrows on trees, rocks, buildings, streets and other creative places to guide the way. This has now been supplemented with scallop shells and small, stone monuments imprinted with the kilometers remaining to Santiago. It is like an adult treasure hunt. Thanks to Father Valina, a pilgrim barely needs a guidebook except to gauge the distance between villages.

We pass an odorous dairy farm, cross a walking bridge, and with a sigh of relief, I see the trail disappear into a forest. I am enamored with the small stone bridges we traverse over streams in the woods today. The path winds through the trees, and I half expect a giggling fairy chased by a gnome to peek her head out from behind a majestic oak.

The terrain is hilly, weaving in and out of the forest. After a particularly heinous climb, a small clearing in a wooded area beckons us to sit and take off our shoes and rub our feet before continuing on. Then randomly, in the middle of nowhere, we come upon a concrete path bordered by shrubs at head height. It's so quiet and lonely, it's a little creepy. I'm glad to have a walking partner.

Mary Jo and I both sense the creepiness and are now a little on edge as we enter more woods. The trees are closer together and closer to the trail. A sound to the left spooks Mary Jo, who is in front of me. She turns and screams in my face. I scream right back at her. When we realize it's nothing we dissolve into laughter. Simultaneously, the need for a pee break becomes urgent, and we race to the side of the path, pulling down our pants a little too late. This makes us laugh all the more.

Just past the halfway point in Viskarret, an enterprising bar owner has built his business right on the Camino trail. The outdoor patio is already half-filled with pilgrims. The scene is colorful and inviting. It's a no-brainer to stop for lunch. I prop my backpack and trekking poles against a table and go inside to order a salad.

"Con atun?" the busy bartender asks rapidly in reply to my fumbling request for *Ensalada mixta* (mixed salad). Uh-oh, I have no idea what he asked – not wanting to turn this into a Saturday Night Live skit, I decide to pretend I do.

"Si, si." I nod enthusiastically. If it's olives, I can pick them off.

I return outside and plop down next to Mary Jo with a sigh. I look around at this sweet little village, peaceful and glistening in the midday sun. It's tempting to stop here for the night. Yesterday was exhausting. But walking with a group has its downside. Most of them are already at our agreed upon destination. In hindsight, I might have been able to stave off future problems with my feet by going shorter distances now, while my body is becoming accustomed to the new Olympic feats I'm asking of it. But I'm not ready to break away from the pack yet.

"Oh, it's tuna," I say smiling, pleasantly surprised and pleased to have solved this puzzle as he hands my picture-worthy salad across the counter.

Lunch temporarily revives me. Caught up in a timeless zone of pure presence, we continue our walk. Delirious with desire for sleep and a hot bath, Mary Jo and I decide to get a private room for the night. We conjure up a beautiful vision of a room with en-suite bathroom. We remain unaware of a ticking clock that will bring us out of our exhaustion induced euphoria.

According to several resources, our hike should be 22 kilometers today. But it would appear the Spanish have a unique relationship with distances. One random sign in the woods shows 6.5 kilometers to Zubiri. However, 20 minutes later the next signage indicates 7.5 kilometers. Mary Jo has tested the GPS system on her phone and found it to be accurate. According to her phone, we walk five kilometers more than we were supposed to today. Not much to do about it but continue walking and keep a sense of humor.

The plan for today was to go on to Larrasoaña. We enter the beautiful town of Zubiri at 4:30 p.m. It is several kilometers shy of our goal, but we are trashed. I stop on the arched, stone bridge and look down to see a few pilgrims relaxing on the bank of the meandering river. I smile. That will be us soon.

The town quickly swallows us up, with its tall buildings. Blacktop replaces the charming cobblestone streets. Crowds of people, many are pilgrims, mill around us. I can't imagine where they all came from. We didn't see many on the trail today. Oh dear. This is a rude awakening after our blissful walk in nature all day. I start to get a little spacey as I feel myself leaving my body.

Survival mode kicks in, and we begin to look for beds. As we peer in the window of an albergue, a fellow pilgrim standing next to us says, "You won't find a bed in this town tonight. It's *completo*. Fully booked."

"Huh?" *No comprende!* "What do you mean there are no beds?" I repeat stupidly as if I don't understand English.

"You know it's a holiday weekend, right?" he adds.

Is it a weekend? I have no clue what day it is. I've completely lost track of time. Let alone keeping track of Spanish holidays.

"Well, we'll have to get a hotel room then," I say to Mary Jo, she stares at me blankly.

"Nope," he says. "Hotels are gone too." He's the tiniest bit smug. I'm guessing *he* has a bed and in some perverse way is enjoying being the bearer of bad news.

"Well, where are we supposed to sleep?" I ask naïvely as if it's his job to find us a place.

He shrugs his shoulders, "Larrasoaña is *completo* too. Sorry. Good luck!" He turns and walks away. There's that damn word again, *completo*!

Ok, so there is no room at the Inn. Mary Jo still hasn't said anything, I think we're both in a little shock. My practical Taurus nature fights to regain control of this unforeseen situation and pulls me back into my body to solve this survival problem.

We continue to wander aimlessly towards the next intersection. Neither of us knows our next move, but Mary Jo's face lights up as she spots Pamela. She's at an outside table with another pilgrim, having beer and wine and laughing. Pamela raises her hand in greeting and waves us over.

"Oh, I want a beer!" Mary Jo exclaims. While I have indeed heard her talk at length for the past hour about her desire for a beer, I can hardly believe she could actually enjoy one now. But that is precisely what she intends to do. With a big smile, she abandons me and joins them at the table.

I jump as a bell clangs behind me, loud enough to drown out all the noise and traffic. It continues on four more times, vibrating through my body, announcing it is 5 o'clock. I turn to see an old church with a tower sporting a massive bell across the street from the bar. I'm awake now.

I follow Mary Jo over to the table. I tell her I will go find us a room for the night. She smiles distractedly, and without looking at me, says "ok, good idea." Pamela erratically moves her arm and knocks over a full glass of red wine, spilling some of it on my $90 merino wool shirt, one of only two I have brought for this journey. I now have a permanent momento of Zubiri. They all laugh loudly as she apologizes and tries to ineffectually mop the table with two napkins, leaving me to figure out my shirt. Obviously, Pamela has been here for a while. Mary Jo pours herself a glass of wine from the bottle on the table before ordering a beer. She dismisses me with a cheery, "see what you can find."

As I walk away, I am not too tired to notice resentment. Oh, I get it. She can have a drink and relax because Angela will do the responsible thing and find us a place to sleep. She was quick to get my number. Mama the care-taking bear, a familiar role. That's another layer of persona I'm ready to shed on this trip.

My self-righteous martyr has a moment of glory before I admit that it only takes one of us to find a room. *Maybe it'll be her turn another night*, I think optimistically. Honestly, I'm too tired to deal with emotional turmoil. Denial uses much less energy.

Walking along I see our friend David. He is shuffling slowly, shoulders slumped, and in a daze. He tells me he has just plunked 2 euros (approximately $2) down for the privilege of sleeping on a gymnasium floor. The last alternative for desperate pilgrims needing to get out of the cold for the night. He leads me to the gym, and I pony up 4 euros to secure two spots for Mary Jo and me. I invite David to follow me back to join us for some liquid courage to get through the night.

The church bell chimes once to announce the half hour. *Hmm, I wonder what time of night the bells stop chiming?* I naïvely think to myself. *Surely, they will stop by 9 or 10 pm so we can get some sleep.*

I'm typically a cheap date. With just a salad for lunch and an uncertain fate awaiting me tonight, it takes only a couple of glasses of wine, and I'm drunk for the first time in many years. Suddenly homesick and in giddy denial, I drunk-dial my three daughters. Laughing and talking loudly to be heard over the traffic and revelry, I tell them of my adventure so far. I find out later that they call each other immediately afterwards. "Is mom drunk?" Michelle asks with disbelief. Alicia laughs, "She's just having the college experience she missed out on because she got married and had us instead."

We claim the last reservations for pilgrim dinner at 8:30 p.m. But by that time, I have bypassed hunger and am coming down from the alcohol high. All I want is to sleep. The food and the company are forgettable.

Throughout the evening, I notice David and Mary Jo are hitting it off. Is it more exciting because they are both married and away from home? It's late as the three of us leave dinner and walk under the streetlights towards the gym. I lead the way as Mary Jo and David giggle and laugh a few steps behind me.

I've had my share of unfaithful men, and the memory triggers a familiar unease in my belly. My respect for Mary Jo is dwindling. I reflect on the hypocrisy of her insisting that our group was women only, but at the first opportunity, she's acting like a teenager drunk on hormones. I am angry, judgmental, and confused at this curveball I've been thrown. This is not what I was expecting to deal with on the Camino. But I'm too exhausted to process it now, so it's duly-noted for future reflection.

The church bell chimes ten times. In the stillness of the dark night, with no other competition, the sound is amplified. It is loud enough to be heard by all in the town. It is late by Camino standards. The door of the gym creaks as we try to enter the cavernous space quietly. It's freezing in here! We are greeted by snoring from several directions. We walk in the dark to the opposite wall to set up camp. The floor is concrete. The 2-euro admission fee did not include mats. We lay our ponchos down under our sleeping bags in an attempt to ward off the cold.

As I slide into my sleeping bag, the unyielding, and icy concrete floor lets me know there will be nothing resembling comfort tonight. Thank goodness I have a small, blow-up pillow. But I will leave in the morning with blue bruises on my hips.

Over wine, the three of us joked about spooning to stay warm. But David and Mary Jo in tacit agreement, use one sleeping bag on the bottom to both lie on, and one on top to cover them. They curl into each other. I am on the other side of David. In desperation, I get as close as I feel comfortable, without touching, to his radiating body heat. He occasionally reaches back and squeezes my hip. It seems like an invitation to get closer, an attempt to include me and help me stay warm. But it feels too intimate, and I keep a distance. The cold, concrete floor wins out, sucking the heat from me like aluminum seats in a stadium in winter. My body shivers and teeth chatter as I toss and turn.

Clang. Clang. Clang. Do we *really* need to know it's 3 o'clock in the morning? But the bells break the monotony of the long sleepless night. I have the rhythm down and count along with the clangs. It saves me checking my phone to see if it's morning yet.

Day 4: Villava

I cannot say I woke up, as I didn't exactly go to sleep. The single clang announces its 4:30 a.m. I verify it on my phone and roll over to whisper to Mary Jo and David, "Are you ready to start walking?" I heard David snoring quietly a few times last night, but I suspect neither of them got much sleep either.

The old gymnasiums small, co-ed locker room is not glamorous, but the hot shower feels fantastic. We stow our sleeping bags and last night's surreal experience and prepare to leave. As we exit the Gym, I confidently turn left onto the main street. David stops. "Maybe we should go back to where we entered the town last night and look for arrows from there." I hadn't thought of that. It turns out, the Camino bypasses Zubiri and doesn't actually go through it. I mentally take note of this strategy for future use.

I am the only one with a headlamp, and I use that term loosely. It is the size of a silver dollar and clips on the visor of my hat. Nonetheless, we are grateful for it. We step out of the street lighting into a dark, narrow lane consisting of large slippery rocks. We carefully pick our way from one to another in the semi-darkness.

The very act of walking revives our spirits. The lane opens up to a road, as tenuous daylight births a new day. We watch as the quarter moon sinks behind the mountains on our left and the sun begins its rise over the mountains on our right. Several kilometers later, we are once again in the lovely Spanish countryside. In unspoken agreement, David has now joined Mary Jo and me.

They were not serving breakfast at the gym this morning, so we are on the lookout for the next village, five kilometers away. It is still early as we enter the empty streets of Larrasoaña, not a soul in sight. My heart sinks at the thought of having to go further before I get coffee on this of all mornings. Finally, we see an old Spanish woman standing on her porch conversing with her neighbor, out for a morning walk. "Hola!" says David with a wave. He asks in Spanish where we might find a café con leche. He doesn't use any sign language, I'm impressed.

The tiny *tienda* (store) at the other end of town is open. Our gratitude knows no bounds. The owner wears a big smile and a halo, as he serves us our first, delicious café con leche of the day with the requisite piece of toast and strawberry jam. This appears to be all the Spanish eat for breakfast. Butter on the toast seems to be an American custom and must be requested additionally. Unfortunately, the Spanish words for butter (mantequilla) and green tea (manzanilla) sound confusingly similar to me, so several times during my trip I'm served green tea instead of butter with my toast.

The walk continues on country lanes meandering in and out of pine forests and large fields. We bypass villages we can see in the distance. We stop to pet some horses. The word *bucolic* comes to mind. As we pass through the woods, I observe that they are not dense and impenetrable like Hawaiian tropical forests. Instead, they are reminiscent of the woods of my childhood in England, with enough space between trees to allow sunlight in and short ground cover to grow.

We overtake two women shouldering full backpacks walking slowly. Their similar features suggest they're related. The older woman is short and rotund and appears to be the mom. Her cropped hair frames a pretty face. The exertion of carrying excess body weight on this grueling journey shows on her face, she appears to be struggling. The younger, teenage girl we assume is her daughter. She has long hair and the same pretty face and is effortlessly keeping pace with mom. "Buen Camino," we say as we pass, the standard greeting that joins all pilgrims in solidarity.

We follow a river as calm and peaceful as our mood, under a cloudless sky. A one-lane bridge with green railings crosses the mirror-like water and leads to a sweet oasis. The bar beckons us with the welcome sight of pilgrims eating and sunning themselves on the grassy bank. It's a no-brainer to stop here for second breakfast. I choose a *tortilla,* which is a crust-less quiche usually made with potatoes, to go with my café con leche. It's a more substantial offering than the *tostada* (toast) we had earlier. Deidre and Camilla are here with their new walking partners. Pamela has walked on ahead. We chat briefly. I will not see them again until Santiago, but we will remain in touch daily via text on WhatsApp.

Over breakfast, Mary Jo, David and I make a unanimous decision to book at a hotel for two nights in Pamplona. Mary Jo and I will share a room. We need to lick our wounds from last night and recuperate.

After several kilometers of more ascents and descents, we cross a highway and are confronted with a flight of steep stairs going up 200 feet. What happened to the level, easy terrain depicted for today in the travel guide? Each stair is four feet long. It takes me a while to get into a rhythm, but the length of the stair adds two flat steps before the next step up, giving some relief. I allow the cadence to take me into a semi-meditative state, which distracts me from the difficulty. I am able to complete it without a break. My legs are wobbly when I finally reach the top, so I gratefully stop to wait for Mary Jo and David.

Several easier kilometers later, we have stopped to admire the charming old buildings that serve as a backdrop to a picturesque Roman bridge crossing a wide river. We think we have reached Pamplona. Our joy is short-lived, however, as we discover we are looking at Villava, a suburb of Pamplona whose town center is another two kilometers. We realize belatedly the hotel we have booked is in Villava, not Pamplona. Under normal circumstances, re-booking a room and walking a little while longer would be no big deal. But in our exhausted state, it feels monumental. Our hotel, it turns out, is two kilometers off the Camino and the slog to get there is through a busy, commercial part of town. It would have been easier to go on to Pamplona. Fortunately, the hotel is modern and inviting.

We are barely settled into our room when Mary Jo informs me she wants to meet David in the hotel bar for a drink. It would appear that they had discussed this while I was walking ahead. I feel the chasm between us widening, the easy camaraderie of the first two days is long gone. Honestly though, I can think of nothing but the bathtub, this means I can take my time. For now, it is enough for me to not have to walk by myself. I'm reminded of the immortal words of Scarlett O'Hara; *I'll think about it tomorrow.*

I dump the whole bottle of hotel shampoo in the water to create a bubble bath. The decadence of soaking in hot water is in direct opposition to last night's fiasco and feels a little un-pilgrimy. But I get over it quickly, mentally grateful that I didn't make myself hard and fast rules about having to tough it out in communal albergues the entire way. Ah, but if that's true, where did the momentary guilt come from?

I actually thought I came on this pilgrimage with no expectations, open to what the experience had to offer. Hah! I'm quickly being shown that I indeed came with lots of hidden agendas. Although in my defense, having such a dichotomy of experiences between yesterday and today is a little unsettling, and has thrown me off my center.

For this evening we decide to forgo a pilgrim dinner and go for the real deal. Bad news, the hotel restaurant doesn't open until 8:30 p.m. Good news, the meal is fabulous. The Spanish timetable is going to take some getting used to.

On the 5th day, she rested...

I am hesitant to disappoint Mary Jo, I know she is looking forward to exploring Pamplona today, but I really need a day to myself and to rest my feet. I ask if she minds if I spend the day here. A little too quickly and with her big smile, she says, "sure, no problem, I'll just go with David." And like that, within a few minutes, she has her boots laced up and is out the door. Hmm, was that too easy?

Life on the Camino sometimes feels like 52 card-pick-up. Every single routine, habit, and way of doing your life has to change. I started out thinking: *I have to have a cup of coffee before I start my day – I need more than two sets of clothes for 7 weeks – I don't eat white bread – I could never sleep on a cold concrete floor for an entire night.* I realize that in the comfortable confines of the life I have created for myself, I have an illusion of being in control. There! I said it. But when I step outside of those confines, I lose control and have to surrender to circumstances.

On the mundane level today, I catch up on my laundry, which means I hand wash everything in the sink of the hotel room and hang it around the bathroom. I go for a lovely walk into the old town of Villava in search of a bank, a Pharmacia and a bakery. Now that I've started eating bread, I can't seem to stop. I have eaten more of that gluten-filled carbohydrate in the last week than I have had in at least three months, it is served at every meal. *When in Rome do as the Romans,* my mother used to say.

Everywhere I go, I have to ask for directions to the next place I want to find. Each time, the person abandons their cash register and other customers, to escort me outside the door of their store, point in the direction they want me to go and rattle off something in rapid

Spanish. I somehow understand the universal language of hands waving and fingers pointing.

My life here, seems to be a surreal study in extremes. Two nights ago, I slept on a concrete floor in a sleeping bag. This evening, we are to eat at a five-star restaurant. On their way home today, David and Mary Jo asked the cab driver to recommend a nice place for dinner, and he sent us to one of the finest restaurants in town. Had I known, I would have cleaned the mud off my boots. It seems that the extremes may be my angels trying to shake things up a little, loosen up all the ways I'm stuck.

Day 6: Uterga

The weather Gods are being very kind to us. We awaken to another beautiful day, and I am ready to walk. Since David and Mary Jo walked around Pamplona yesterday, and I did over five kilometers myself, we decide to take a taxi into the city and leave from there. It was surprisingly easy to justify busting my *taxi cherry* and blowing up another of my Camino rules.

We hike out of the beautiful old city of Pamplona, past the University, stopping in Cizur Menor for coffee. Shortly after, we start the climb up into the surrounding hills. The path rising above us, going higher and higher, is bordered by a random collection of colorful wildflowers with red poppies predominating. The fields beyond the border contain yellow rapeseed flowers as far as the eye can see, punctuated with occasional, rogue poppies. The sight lifts my spirits, every cell in my body feels alive.

My body is getting stronger each day, but this is a steady uphill climb for several kilometers. I keep my eyes on the prize; the windmills positioned along the top of the ridge. This hilltop called Alto del Perdón is considered a place of forgiveness. I know from many pictures of this landmark that there will be a series of metal statues of pilgrims situated next to the row of windmills. However, the statues will not be visible until we reach the top.

Halfway up the hill is the welcome sight of a small bar. Pilgrims in their colorful gear, lounge on green plastic chairs and sit on the ground, propped up against a wall. It's the first building after the church, in a small picturesque village, contained within one street. The white stone of the buildings and road are in sharp contrast to the deep blue sky. The effect is almost blinding in the midday sun.

Four people standing in line fills the store. They have an equal amount of space behind the counter for two people hurriedly filling the lunch orders. My tortilla is bland, I am underwhelmed by a dish I have heard so much about, but who really cares about such mundane things on an adventure like this.

Now the going gets tougher. We continue to climb on ever steeper and narrower trails, winding their way through rocks and scrub brush. We are wowed by views of the snow-capped Pyrenees, that we conquered several days ago. The steady drone of a getting-to-know-you, conversation between David and Mary Jo, is starting to grate on me. I want to immerse myself in the beauty and silence of nature, and the totality of this experience. I pick up my speed a little to put some distance between us. The trail is rocky, and my knee starts to hurt badly. I'm limping and I'm afraid I will not be able to make this tough climb, so I start repeating to myself over and over in my head: *my knee heals itself and feels better with every step.* Within minutes the pain is gone, and I can continue climbing.

From the top, we survey the magnificence of what we have just traversed. We take pictures with the figures on the ridge, hang out for a while, then begin the long, steep, rocky descent down the other side. Mary Jo is no longer even pretending that she wants to walk or talk with me, she and David are in their own little bubble. This isn't my first rodeo, I see where this is leading. For the first time, I think to myself: *I'm going to be walking on my own soon.* In retrospect, I'm surprised that I clung for so long to the illusion of having a walking partner, when Mary Jo is clearly telling me she'd rather walk with someone else. It's not like me to hang out where I'm not wanted, but that is the level of my fear of going solo.

Before entering Uterga, our village for the night, we come upon a large statue of Mary. She appears to have a significant presence on this trail. Her protective, mothering energy makes this feel like the perfect place to pray for Cooper today. He is on my mind every day when I look at the profusion of wildflowers everywhere. At 6 years old, he has an unusual passion for flowers, and an artistic gift for re-creating them. He loves to paint, and draw, and fashion clay images of flowers. I take him for nature strolls in areas that at first glance I think: *Oh, too bad there are no flowers here for him.* But by the end of the walk he has gathered an impressive bunch of "bootiful" wildflowers for his mother. He sees flowers where no one else does. I send him at least one picture a day, of different flowers I find on my trip.

Cooper at 7 years old

We walk down the street of yet another quaint, old village in search of our albergue. We pass window boxes adorning window sills with blooming flowers. All is very quiet the town almost seems deserted. Then we come upon a happenin' scene. The two-story sandstone houses are close together, and this one has a small grassy garden, with several groups of pilgrims enjoying beer, wine, and snacks, laughing and chatting animatedly at tables and benches. Others are washing their clothes in the outdoor sinks on the side of the building. The sign in the front yard tells us we have reached our destination, Hallelujah!

We are assigned bunkbeds upstairs, all three in the same room sleeping 20. David and Mary Jo choose two uppers that face each other a distance from the one I have chosen. I have no choice but to have an upper bunk this evening also. We shower, then join the other pilgrims in the front yard enjoying the beautiful sunshine, for some sangria. Following dinner, I retire to write my blog in the sitting room, while David and Mary Jo decide to go for an after-dinner stroll in the village.

While they are gone, I decide that I don't want to walk with them anymore. It's clear that I'm a third wheel and don't have anything in common with either one. It's time to put on my big-girl panties. I feel a sense of liberation.

When they return an hour later, I pull Mary Jo aside and tell her of my decision. She is not happy. She doesn't say that she enjoys walking with me and is sad to see me go. Instead she seems angry and confused by my choice. She tries to tell me that since she's a psychologist, she's spending all this time with David to help him in a professional capacity with his marital issues. I'm able to restrain a derisive snort as I think to myself, *yeah, whatever!*

We go to turn in for the night, and the lights are out. Ten minutes later as I'm lying in bed, I see David creep into the room. He goes to Mary Jo's top bunk, she leans up on an elbow, and they kiss. They apparently think it's too dark for me to see them. *And there it is! Well, that should help him with his marital problems.*

Day 7: Lorca

This morning I fear I have been blessed with the scourge of the Camino; bedbugs! As I brush my teeth, a woman standing next to me points out two small dots of blood close to each other on the back of my shirt – rats! I lift my shirt and ask her to check for bites. There's nothing visible yet, but I know from my research that they can sometimes take days to appear. So, along with the David and Mary Jo drama, I now have to deal with this. Tonight, I will need to find a washing machine.

I leave the albergue at 7a.m. The sun is rising over the mountain, with streaks of pink in the sky. I have a sense of exhilaration and freedom that I have conquered my fear of walking alone. To be by myself in the dead quiet of the early morning, broken only by the chorus of birdsong is pure heaven. The country lane between fields that eases me into today's walk reminds me of the English countryside I grew up in.

After 30 minutes I have picked up my pace. As I travel through the next village, I see other pilgrims heading out on the trail. Like a casual stalker, I follow them from the other side of the street, keeping my distance, but comforted by their presence.

Spain is not flat! That is an understatement. The uphill today is brutal, and not only in the mountains. I marvel at the architectural feat of building whole villages on the side of a hill. Thank God the scenery is so outrageous, and I am blessed with another beautiful Spring day.

I trudge along a wide gravel path in the woods that winds its way steadily uphill for several kilometers. Rounding a corner, the short but ridiculously-steep *Mother* of a finale reveals itself. I stop and bend over my poles, resting my legs and back for a moment. I would laugh if I had the energy. A few four-lettered words come to mind as I attempt to regroup. Then, in a beautiful act of random kindness, I watch as an old Spanish man helps a pilgrim cyclist who had passed me moments earlier. The cyclist has dismounted and is attempting to push his bike but can go no further. The old man who shames us both with his effortless ability to navigate this hill, reaches over and helps propel the bike to the top. This act of love infuses me with the energy to continue. I start putting one foot in front of the other again.

In the distance, the next village is perched like a fortress on a rise in the landscape. As I get closer, its beauty is stunning. The stone buildings in various shades of beige and white with an occasional splash of color, are 2-4 stories high. A yellow arrow points me up a one-way cobblestone street, that looks and feels like a 45-degree grade. *This is why you don't see obese Spanish people.*

A coffee break and some time to put my feet up would be nice right about now, but it's siesta time, and nothing is open, so I continue walking. My plan today was to continue on for two more villages, and I'm making good time, but the heat is sweltering, and the relentless sun is sapping my strength. I turn a corner and there, thank God, is the first village. My body is exhausted, so at the first albergue, I inquire after a bed and a washing machine. They have both. I make the decision to stay.

I wash everything including my sleeping bag. After hand washing for a week, I contemplate the luxury of this common household fixture. I lean out of the second-story window of my room to hang everything to dry on the rack provided for this purpose. Spain appears to have an aversion to clothes dryers. Old women hanging out of windows with wet laundry is a familiar sight here. All I need now is a babushka.

One of my bunkmates is a Dutch woman around my age named Yoko. I will continue to run into her for the next few weeks. She at first doesn't seem too friendly. I choose not to take this personally. Between cultural differences and the varied, personal reasons we are all here, we learn to give each other space.

After eating a late lunch, I look across the narrow street to see David and Mary Jo checking into a different albergue. Sigh! Well, I guess this was inevitable, the Camino is a small Universe. A nameless feeling more primal even than loneliness engulfs me. I swallow my pride and walk over to say hello. I am grateful for the crumb of community their company offers after a day of walking alone.

They are talking to the overweight woman Rowena, and her daughter Rosa, that we passed in the woods yesterday. After I'm introduced, they invite me to join them all for dinner.

Mary Jo says, "I need to go back to the room for a while to do our wash."

David smiles, puts his arm around Mary Jo and with a smug look says, "She's my Camino wife." Mary Jo looks up at him with a loving smile – I think I just threw up a little bit. I try hard not to snort. Why would anyone want to get *Camino married* if it means you have to hand wash their dirty socks and underwear at the end of the day – no thank you.

But like some weird Twilight Zone episode, I'm happy to go along with pretending nothing has happened between the three of us. I'm not even really sure what my jumble of feelings is. But a compelling need to feel included, appears to override everything else. The Camino can tear you down and bring you to your knees, to face parts of yourself you are unaware of. I realize now this is more complicated than just being afraid to walk alone, as I find I really enjoy being by myself on the trail. It's more about inclusion, feeling part of something bigger, being part of a community.

At dinner, I discover that Rosa is 17, and at 14 she was the youngest girl to ever thru-hike the Appalachian Trail – by herself – wow! I would love to hear more of her story. The few comments she makes show her to be of above average intelligence, and what I would call an Old Soul. But her mother is a firecracker, and the young girl sits quietly tapping on her phone, while we are entertained during dinner by Rowena. We get to hear long-winded tales of *her* experience of her daughter's epic journey, sprinkled liberally with 'F' bombs. Her openness in front of her daughter discussing sex and drugs and swearing like an MTV rapper, shocks me. I cannot tell if I'm being prudish or discerning? I'm not averse to using the well-placed *fuck*, for emphasis and humor. But when it's used as an adjective, verb, and noun all in the same sentence — well, now that I think of it, I guess that takes some skill.

A hot shower, getting the bottom bunk, eating at a table that doesn't wobble, finding what I need in my backpack on the first try without having to empty out the whole damn thing. My life has been reduced to one of exquisite simplicity — like a temporary get-out-of-jail-free card.

Day 8: Villamayor de Montjardin

Today was the day of churches and monasteries. One grand. One austere.

I decide to try traveling with David and Mary Jo again. Maybe decide is the wrong word. One minute we are enjoying a nice dinner, the next we are discussing where to sleep tomorrow night. We seamlessly pick up where we left off as if yesterday never happened. Aaand, it's episode two of the Twilight Zone.

To avoid competing with the general masses for beds (and sleeping on a concrete floor again), we choose to stay at a small village in between the popular towns recommended by *Brierlys*, the Holy Grail of guidebooks for the Camino Frances. There are only four large cities along the 800-kilometer Camino route. The rest are mostly quiet, one-church towns or villages, hundreds of years old and steeped in history. They create the peaceful ambiance many of us are looking for in our self-reflective journeys.

Fueled by a quick café con leche, we leave at 7:15 this morning under cloudy skies. The absence of sun makes for perfect walking temperatures. A meandering little path takes us to and through Villatuerte, another small medieval village. We exit on another trail and are finally in the countryside I'm craving.

I stride out in front of my travel companions to put some distance between us. I'm already wondering why I wanted to walk with them today, their constant chatter is tiresome. I again realize how much I enjoy solitary walking. But at the end of the day, my exhaustion magnifies my vulnerability and feelings of aloneness. My problem is compounded by the fact that I have not yet had a chance to strike up friendships with other pilgrims.

My eye is drawn to some overgrown ruins off to the left. As I get closer, I see that it's a tiny, four-walled, medieval church (1062 ad, it turns out). On the one hand, it looks interesting, but not spectacular. I am in a nice rhythm making good time for the day, so I am about to walk past it. But I cannot stop looking at it and finally succumb to the strong urge to go in. The sign in Spanish says something about an allegiance this church has with Archangel Michael. He is recognized by most, if not all, religions and spiritual traditions as a powerful divine messenger. A healing angel, he is often seen with his sword as the protector and defender against evil forces.

A wave of emotion envelops me and causes a tear to roll down my cheek, as I walk into the tiny room with only two small slits for windows. Ah, this is my place to pray for Cooper today. I guess today he needs a sword-wielding warrior, not just a protective mother (Mary) looking over him. I stand at a stone pulpit in front of the stone altar, with a large wooden cross above it, the sole adornments of this ancient place of worship. As I pray, a vision of Archangel Michael wielding his sword comes to me, and I ask that he cut the connection to my grandson's affliction and that it never revisits him. I envision him slicing down through the energetic cord that attaches Cooper to his illness. Now my tears flow in earnest – I have the sense my prayer is answered. Without waiting for David and Mary Jo, I turn and leave the church, walking on ahead by myself for a while to process this powerful vision and experience.

The three of us approach Estella, a large town and a popular destination for overnight accommodations. I'm admiring the red poppies mixed with purple and yellow wildflowers, peeking above the tall grass on the side of the path, when suddenly this beauty is upstaged by a noxious smell. We pass three large, white tanks that I later discover is a pig farm.

I fall in step with David and Mary Jo as we leave the odor, strong enough to kill any ideas of lunch, behind and enter the town. A majestic church looms above a 50-step staircase on our left. For a moment, all I can see are the steps, and I shake my head and groan, "no," as I roll my eyes. I'm tired, and my feet really hurt. But this place of worship is calling me also. David and Mary Jo are already part way up as I find my meditative space and start climbing.

The sound of soft, heavenly chanting as we enter is transporting. Where is it coming from? I can see no one by the altar. As I pan around, I realize that six pilgrims are sitting in the pews quietly serenading us. We walk around admiring the stunning gold leaf architecture and stained-glass windows. Part of walking the Camino is like going to a museum every day. In a little alcove, I stop and smile at the statue of Mary holding her Child, that I am starting to feel such an affinity to. She is everywhere, seeming to even take precedence over Jesus in many of the churches. I ask her to watch over Cooper, then light a candle for him.

In one source, I read that when the Christians came to infiltrate Northern Spain, there was at the time a Goddess-worshipping religion here. In a diabolical stroke of genius, it's thought that they may have put Mary forward in a place of eminence, hoping that the people would more easily be able to relate to Catholicism. Of course, all pre-Christian records have been destroyed, so it is difficult to verify. Even though the *Holy Grail*, somewhat-controversial myth, of St. James takes full credit for the Camino's existence, several sources claim it was a pilgrimage long before Christianity set foot in Northern Spain.

Following our church tour, we walk through the busy town without stopping and approach one of the most famous attractions on the Camino, the free wine fountain at the Irache Monastery. I arrive before my companions and remove my pack to take off my scallop shell. A large, three inch in diameter, scallop shell is a symbol of the Camino, tied in with the legend of St. James. These are sold at the pilgrim office and worn by most pilgrims attached to the back of their pack. It is traditional to use this to receive a taste of the wine. I hold my shell under the spigot and the wine trickles into it. I smile and have the thought that after the experiences I've had in the churches today, it feels like I am taking communion. All I need is a piece of bread to go with it, getting that later won't be a problem.

I notice a man and woman standing nearby with a big camera set-up. They ask if they can film and interview me drinking some wine. "Of course." They tell me they are with Basque TV. I will miss my five minutes of fame however, as I will not be near a TV for weeks.

With the memory of the pig farm receding, we decide to have lunch at a large deserted modern bar in the next small village. It is only us, in a place big enough to hold a wedding reception. As we are waiting for our food, we pull out our phones to look for a place to sleep tonight. Aaand the weirdness is back! It is evident that David and Mary Jo now want to be the ones to share a room. Since we are looking at private hostels, that means I will have to pay full price for mine. I'm feeling decidedly irritated by this whole situation, mostly at myself for wimping out and allowing this to be prolonged. The analogy comes to mind, *"fool me once ..."*

We finally all agree upon a room with three beds, in what's called a Casa Rural (country house/private home). Back on the trail, I silently berate myself for not having the courage to break it off with them. I walk out in front and allow Mother Nature to pull me into her embrace. The path continues through woods and countryside. Then finally, the *Hallelujah moment.* A church spire, which by now we've come to associate with the next village, rises tantalizingly in the distance at the top of a three-kilometer long hill. This will be our home for the night, so I don't give myself time to think about the walk ahead of me or to feel sorry for myself. I have learned to focus on the natural beauty around me and not on my pain. It starts lightly raining halfway up. We've had such great weather so far, I'm not going to complain about this either.

We are welcomed into a lovely home, the low ceilings and thick, clunky trim betrays its age. The bonus plan tonight, we have our own private bathroom, quite luxurious by Camino standards. After a shower and short self-massage, my feet and legs feel the best they have for a week. The blisters on my baby toes are well on their way to healing, and my knees are good too. Still, I stagger a little on my way to dinner. An unrelenting, non-specific pain has started to plague my feet.

We go to a local café. I say that as if we have a choice. Let me re-phrase; we go to the only bar in the tiny town at the top of a mountain we are staying at, and we are grateful to get the last three seats in the dining area.

The room is already crowded, so we go to the bar for a glass of wine while we wait to be seated. A man, who turns out to be from Switzerland, is standing next to me. I later dub him Bond – James Bond. Mary Jo and I decide he is a cross between Bond and the Marlboro man, handsome and rugged, with gentle lines on his face etched by time, though probably a bit younger than me. No doubt he is a ladies' man, I recognize the overconfidence, *the eye* and the grin he gives me – *Wait, what? Is he looking at me?* I casually look over my shoulder to make sure there isn't some gorgeous, blonde, 30-something standing behind me – nope. Hmmph, well whaddya know? I smile back and watch a slew of conflicting thoughts jockey for position in my mind. I honestly can't remember the last time I flirted with a man. Oh, what the heck, they haven't invited us to sit down yet.

So, for 10 minutes before dinner, and 15 minutes after, I enjoy a mild flirtation with this handsome man. This consists mostly of me smiling and nodding with one-word responses, as he speaks limited English. His friend sits quietly by, looking a little bored. My guess – he's used to being Bond's sidekick.

Day 9: San Sol

A cold, windy, drizzly day greets us this morning. We dress in layers and rain gear. I don't mind the change-up in the weather, it's been amazing so far. The problem with ponchos is that they don't breathe, so I begin sweating immediately. The sweat evaporates and causes condensation on the inside of my poncho, which then gets re-absorbed into my clothing. All the layers quickly become soaked.

The three of us weave our way through the tiny village we stayed in last night until we see a yellow arrow directing us to a path leading down the mountain. Pruned back grape vines on the left, are beginning to show signs of life after their winter hibernation. On the right, emerald green wheat fields wave a lush good morning. In the distance, we see a village, but it's too soon to tell if that's where our first cup of real café con leche will be.

The meal that came with our room this morning consisted of brewed coffee (my inner coffee snob was twitching), a box of orange juice and Wonder-Bread-type toast with butter and jam. We have a long 13-kilometer walk this morning, to get to the first town where we can stop for second breakfast. So, we had to lower our standards and be grateful that there was at least some food for us.

I am bewitched by the bridges we encounter every day, crossing small streams. They vary in size, architecture, and age. Some, I know, are part of the Roman road system and very ancient. The landscape, in all directions, is like an artist's palette in every shade of green. We are frequently passing or being passed by other pilgrims, so when I hear someone loudly call, "Angeleeeena," I turn to see Bond grinning widely at me. He and his friend are bearing down on us at an impressive pace. As with last night, a conversation is awkward, so after a brief greeting, they walk on. I'm sure that is the last I will see of my Marlboro man, but it puts a spring in my step for a while. My flirty interaction with him last night brought up feelings and questions I hadn't asked myself in a while. *Do I want another relationship? Am I up for the fun but exhausting process of flirting and dating again? I could get used to this undeniable warmth spreading from my gut to my toes and fingertips. It sure is nice to feel desirable again.*

The wheatgrass growing in endless fields on either side of us is 18 inches tall and waves gently in the breeze, soft as a woman's hair. We pass a pine forest on one side of the trail, and some random bushes dripping with tiny pink blossoms on the other. These micro-moments of intense beauty fuel my soul.

The path is wet and muddy, and there is no sign of any kind of habitation, (read: second breakfast) for almost four hours. Stopping to pee with all this gear on in the rain is challenging but necessary. I have learned to make sure there are no jackets tied around my waist when a pit stop is necessary. Between the drizzle and the mud, we take very few rest breaks as there is nowhere to sit. It is raining hard enough that I have to keep my poncho on, but the heat generated by the walking causes me to continue sweating and dripping condensation inside it. Frankly, I might as well take it off and walk in the rain.

I'm a soggy mess as we go around a corner on the trail to see our destination looming ahead ten minutes away. Los Arcos is quite large. Its town square is overshadowed by an enormous church, whose doors are locked. We wait in line at a busy bar. My mouth is watering for fresh orange juice, café con leche, and tortilla with a piece of freshly baked bread.

It has finally stopped raining, and I am able to remove my poncho for the remaining eight kilometers, much of it on a one-lane country road. I catch the first daunting glimpse of our destination far in the distance. I burst into song, "Hallelujah!"

We are once again booked into a lovely Casa Rural. I am noticing that David, a successful businessman used to the finer things in life, isn't planning on roughing it any more than he has to on this trip. He prefers better accommodations and whenever possible he foregoes the pilgrim meals. I think this works for Mary Jo also, although I can't be sure, as she is definitely letting David take the lead. In today's world there are many options for completing this journey, and without walking in another persons' shoes, we can't know their history, possible medical issues, or motivations for this trip. So, I feel the need to be generous of spirit regarding others and their choices. Besides, by the end of this pilgrimage, I find myself gravitating towards nicer accommodations as my body begins to break down. But for now, I want to experience all this undertaking has to offer, including the large, noisy albergues and all their quirks. For me, it's all part of the adventure. This will contribute to my decision later on today, to make a go of it alone again.

One night at this lovely place, with dinner and breakfast, is 35 euros each. Removing your boots at the door is mandatory in most places, so we walk up three flights of stairs in our socks. The last flight is highly polished stone, aka slippery, and on a dangerous downward slant. I make a mental note to be very conscious walking up and down stairs today. I take one of the four single-beds in the large attic room we have reserved. David and Mary Jo take the two in the far corner.

When we have all completed our end-of-the-day rituals in preparation for the next day, David crawls into Mary Jo's single-bed with her under the covers. This is a superhuman feat, as he is not a small man. In one hand he holds his phone, scrolling with his thumb asking her which movie she'd like to watch. Mary Jo catches me looking at them in disbelief, "Do you want to join us?" she asks. I don't think this is a kinky request, so I can only assume she's being sarcastic. She certainly can't be serious!

I somehow manage to keep a straight face as I reply, "That's ok thanks, they have a washing machine, I'm going to take advantage of it." What I really want to say is, *"The hell with this shit, I'm outta here!"* I realize I have witnessed the last straw. Being the third wheel in what appears to be a burgeoning affair, is not what I signed up for. Time to stop sucking my thumb and leave on my own tomorrow.

Day 10: Logroño

Breakfast in the otherwise deserted kitchen is a quiet event, as if Mary Jo and David know the decision I have made. There's nothing left to say. Their *business* (which indeed it is) brings up feelings and memories from my second marriage to an alley cat of a man. Even my first husband, before he was that, and we were in the heat of our teenage, passionate love for each other, thought it would be ok if he had an affair as long as he confessed to me about it. Crying declarations of, "I love you so much," "She means nothing to me," "I'll never do it again." - *Is this what it looks like when you 'love me so much'?* Well, nothing says: *Fuck you!* like taking all the money I had been saving towards our upcoming year of newly wedded bliss and buying a horse with it. To my knowledge, he never cheated on me again – I won that battle. His saving grace was that he was an excellent father, but without the proper tools to navigate the passionate, turbulent marriage to my high school sweetheart, we both eventually lost the war.

So, when enough is enough, I do what I do. I am firm in my resolve to walk off on my own today and not look back. From the second floor of the Casa Rural, framed by the window like a painting in a museum, I look out to see the first village I will be heading towards this morning, only a kilometer away. The red roof of the house below ours is almost obscured by the tops of trees. Beyond the buildings in the distance are wheat fields, a solitary tree, and green hills. Like looking into the eyes of a newborn child, the perfection momentarily takes my breath away.

I slip downstairs by myself while David and Mary Jo finish getting ready. I begin my walk in a light drizzle with heavy wind. The dead quiet of the morning contributes to a familiar sense of peace, bordering-on euphoria, that comes from taking back the reins of my life. I know I will have to temper the euphoria though, as too much of that can lead to an unaffordable crash later, and I need to stay the course this time. With no intermission to herald it in, I feel like Part Two of my Camino has begun.

Today the wind has a voice, a deep, haunting whistle, like a friendly ghost, causing the bushes to wave at me as I pass. I have never really liked wind as it makes a mess of my hair. But I'm finding that I have a different relationship with the elements on this trip. I am accepting them and adjusting, rather than thinking that Mother Nature has to bend her will to mine.

As I walk entirely on my own, with not another pilgrim in sight, my thoughts track inward. I realize that my Pollyanna view on life wants experiences to be picture perfect. Instead of seeing Life as this beautiful tapestry of yin and yang, I try to delete the events that don't fit the ideal, pretending that they didn't happen, or that I've healed and moved on. In hindsight, I will look back and see the importance of these last few days. I will be reminded of the emotional turmoil embedded in the DNA of my heart, by men I have loved and trusted. I have often found that any healing process first requires some conscious recall. Mary Jo and David were kind enough to evoke that.

After the first village is a beautiful broad path. It is laid with pinkish, modern geometrical stones, bordered on one side by tall shrubs waving a welcoming greeting in the wind. After an hour on the trail, a quick glance behind me confirms Mary Jo and David are nowhere to be seen. I'm sure they will be glad to be rid of me.

I'm a lean, mean walking machine this morning. I traverse the long, steep hills with sparse vegetation, at a steady pace without losing my breath. I notice how important it is to keep a rhythm and a strong mental attitude. Any focus on discomfort becomes a downward spiral, turning my legs to lead and my mood to despair.

In Viana, I stop for second breakfast at a noisy, crowded bar. A man and woman I met last night at the Casa Rural, invite me to join them. Mary Jo and David come through the door as I'm finishing. Chairs are at a premium, so the couple offer them theirs as they get up to go. Oh dear, this is awkward! Our greeting is cordial, and our desire to allow this walking partnership to dissolve feels mutual and strangely unemotional. I leave while they wait for coffee. One more brief meeting a week from now will give me a final insight into our relationship, along with a sense of closure.

The rain holds off for the remaining ten kilometers to Logroño. But the approach on concrete into the city is grueling, and my energy is waning fast. As I approach a sweet-looking young man, he turns and says, "Hello, where did you start from?" His English is good, but I detect an accent.

"St. Jean, and you?"

"Oh, I started from my front door in the Netherlands, I've already walked 1,000 kilometers." He says humbly.

"Whoa, that's impressive!"

He tells me his wife and disabled son are cheering him on from home. He is not looking for sympathy, as he matter-of-factly tells me the heart opening story of the tough hand he and the love of his life were dealt with their son. However, it is evident from his sharing, that part of his Camino is to somehow sort through it all. If we ever did exchange names, it wasn't my takeaway from this single encounter.

This happens quite frequently on the trail. Just when I am wondering how I could ever walk the last couple of kilometers, my angels will send along someone to distract me. Mr. Netherlands will also hold the distinction of being the first of several men on this journey, to display unconditional kindness, gentleness, and thoughtfulness. This begins the healing of a heart broken too many times in my complicated relationships with men.

We finally walk over a bridge, busy with two-way traffic and people. On the other side is a Pilgrim Assistance center. Logroño is a large city, and I'm not up for wandering aimlessly for an hour. So, I forgo my usual M.O. of: *get lost first, ask for directions later.* We both decide to stop in to get maps personalized with red lines, arrows, and circles to guide us to our respective places for the night.

After I part ways with Mr. Netherlands at a busy traffic round-about, I obsessively check the map to make sure I'm going in the right direction. I was done with this day of walking five kilometers ago and would rather not spend any extra steps getting to my bed.

I seem to be following a 30-something, young man. I can tell he's a pilgrim by his backpack and dragging feet. But with his neat, clean clothes and gelled back, black hair he could be on his way to read the 6 o'clock news. I'm in awe that he still has enough energy to care what he looks like. Sure enough, we end up at the same place and are assigned bottom bunks across from each other.

"So where are you from?" I ask.

"Colorado," he continues to unpack without looking at me, appearing shy, not rude.

A bird-like Spanish woman interrupts to shows us around. Efficiently she flits from one amenity to the other, explaining *the ropes* in sing-song, heavily accented English. Camino-mama insists we come to sit in the kitchen and serves us a cup of instant chicken noodle soup. I smile to myself at the depths to which my culinary standards have slipped. I politely sip at the salty concoction, not wishing to offend her kindness and good intentions. I will not know Matthews name for several hours, he sits opposite me, a man of few words.

I return to my bunk to prepare my bedding and empty my pack to look for something cleanish to wear. It no longer seems imperative to wash everything daily. If it passes the sniff test and still looks clean, chances are it will do for another day. I now have a third ziplock bag with *optional* dirty clothes. The likelihood of them getting laundered depends on the availability of laundry facilities and my energy level at the end of the day. This falls into the *reality rules* category when you are traveling with only two sets of clothing. Sometimes my exhaustion after walking for six hours makes even small tasks, like hand washing, feel insurmountable. Clean underwear however, is never optional, those along with my liner socks get washed daily regardless.

I grab my toiletries and take off to deal with the communal shower scene. There are three showers next to two toilets, everything white, spotless and modern. But the anticipation of shared facilities remains a little daunting. Matthew exits one of the showers wearing only a tiny towel tucked low on his narrow hips. In a flash, I take in his impressive physique. We nod and look away quickly. Why this feels more intimate than being at the beach, I have no idea.

I have the energy to do wash today, so even my *optional* bag goes in. The lines outside to dry clothes on are in a courtyard enclosed by four buildings. The only way in and out is through a door that locks automatically, so I carefully wedge it open and keep a nervous eye on it as I quickly hang my items.

In bunk beds on the other side of the room are two Australian couples around my age. We met initially at Orisson the first night and have run into each other a couple of times since. In spite of being a close-knit group, they are open and friendly to me and invite me to dinner with them. They are planning on going to a more upscale place. I'm still reeling from the days walking and not ready to make plans yet, "That's so kind, thank you, I'll let you know," I say.

I return to my bunk to lay down for a while. I have barely shut my eyes when I hear a loud, coarse, British accent, "Oh Lord, that was brutal." I open one eye to see a short, but surprisingly large woman throw herself, unceremoniously, on a bottom bunk. Her swollen feet and ankles don't look like they can effectively hold the weight of her body. Her cropped hair is the startling color of *fireball red*, it turns out to match her personality. Because Reds' voice has one volume, LOUD! I'm able to deduce that she met her travel mate from Denmark on the trail. Danishwoman is quiet and tall, with long blonde hair topping off her athletic body. A stranger pairing you could not imagine.

I stop pretending I'm sleeping and open my eyes to watch the show. Red is comic relief, her British sense of humor has me laughing out loud, even if it does seem a little over the top, as she laughs raucously at her own jokes. I detect a little uneasiness between the two of them though. Red is obviously in a lot of trouble with her feet, and Danishwoman could leap tall buildings in a single stride. Danishwoman is vocalizing her concern about not having enough time to finish (in small print: *because Red is slowing her down*). But it's clear to see she's torn because of her loyalty towards her new friend.

Matthew eventually sits up also, as sleep is not happening with this sideshow going on. While neither of us contributes to the conversation, at some point, we become included. When they invite both of us to join them for dinner, we say yes. I walk over to politely decline the Australian group's invitation for this evening. Why I'm not sure. The Australian's offer feels safe and comfortable, the other more adventurous, less predictable. Who knows, I just go with my gut.

It is not the last time I will run into the Australians. We hopscotch each other all the way to Santiago, meeting for the last time in the square in front of the Cathedral. We are always polite but never really making a deep connection. However, I sense the potential, some strange familiarity – could we be walking together on the same pilgrimage in a parallel lifetime, with completely different experiences? It reminds me of the power that each decision, no matter how small, has on the trajectory of our life.

"So, Matthew, what brought you to the Camino?" I ask at dinner.

"I sold a successful, legal medical marijuana business, and took a year off to travel around the world."

Well damn! Who knew? His neutral expression is difficult to read. Is this why he seems a little lost? While this idea looks good on paper, I would imagine the reality of spending 12 months with no home base is a bit ungrounding.

We are all a little tired of Pilgrim dinners, so opt for a bar where we can order off the menu. Simple right? The meal is a comedy of errors. First, they don't have the salad I want, so I choose another. The waitress nods as if she understands. But she either doesn't or doesn't care, as I get a completely different salad.

This one consists of pasta on iceberg lettuce and thousand island dressing. The whole thing looks anemic and unappetizing. I move it around my plate for a while, like a kid with broccoli pretending to eat, but I only take a few bites. I end up leaving most of it.

Fortunately, this is yet another day where tired beats hungry. I realize this is the price you sometimes pay for not speaking the language. Speaking of price, that salad cost more than a three-course pilgrim dinner.

Day 11: Ventosa

Unquestioning surrender to what the Camino has in store for me, appears to be the learning for today.

It promises to be another beautiful spring day as I step onto a quiet street at 6:30 a.m. Another lone pilgrim passes me hefting a big pack, nods, and murmurs, "Buen Camino." I return the greeting and fall into step behind him as I retrace my steps from last night. As I approach the bridge I crossed yesterday, more pilgrims start to appear, drawn magnetically from side streets. Like a trail of ants, we all converge then follow each other, quietly looking for that golden ticket, the yellow arrow.

The Old Town magically appears on the other side of a round-a-bout. Modern buildings like the one I stayed in last night, are replaced with ancient stone, two-story structures that house mostly albergues and businesses catering to the Camino. There's a feeling of *coming home,* as I fleetingly wish I'd known to stay here. I quickly remind myself that at least I had a warm, comfortable bed to sleep in last night. I look down to see a bronze scallop shell embedded in the cobblestone, single lane road. My shoulders relax as I'm guided back on the Camino.

The quiet comfort of the Old Town is soon replaced by the bustling, busyness of a large, modern city waking up for another day. Cars jockey for position on streets going in every direction, amidst tall buildings. I'm engulfed in a sea of pedestrians, a mixture of well-dressed Spaniards on their way to work, and travel weary pilgrims looking a little the worse for wear even at this early hour.

A bar invites me in to have a café con leche and croissant. Hopefully, this will sharpen my wits and help me navigate this morning without getting lost. The yellow arrows are not as apparent amongst everything else vying for my attention, and according to conflicting accounts in various sources, there appears to be more than one way to exit this city. As I leave the bar, I stop on the sidewalk to get my bearings. Pilgrims on both sides of the street are heading to my left, so I follow them. Soon I am rewarded with a yellow arrow on a lamppost pointing in the direction I'm going.

My phone tings with an alert. Mary Jo has sent a group message on WhatsApp to the five of us that started together (we all remain in touch this way throughout our journey). The group texts enable Mary Jo and I to communicate indirectly, while keeping a pretense of friendship. She's in Logroño also, staying at a hotel. She doesn't mention David, but I know he must be with her. She has a migraine, so will be sitting today out. While I didn't wish a headache on her, I'm secretly glad she's putting a day of walking between us. It will be easier to avoid them now.

At an intersection, I cross the road. My head is bent over my phone, following a blinking blue dot that directs me to make a right turn to keep going on the main thoroughfare. Honk, honk, honk – I look up, both people in the car driving towards me, eyes wide, are pointing animatedly behind me. I have turned the wrong way. Bless their hearts, the Spanish are so kind and try to help us out whenever possible. Our large backpacks are a dead giveaway. I smile and wave a *gracias* as I turn around and notice other pilgrims in front of me. I had been so intent on my phone app, I hadn't stayed alert to what was going on around me (I smell a learning here). I realized later that the online map would have taken me *through* the busy City, instead of on a lovely path around it, winding through a park.

Two women shouldering equally large packs are right behind me and witness the incident. "Well that was lucky for you," says one with a big smile, in a lilting Irish accent. I fall into step with Eyleen and Karen as we laugh about my near miss. Thankfully, within a quarter of a mile, we find ourselves leaving the town and in a large, well-kept park. These two cheerful friends from Ireland are great company as we talk and walk our way to a scene from a Disney movie. The reservoir is a serene body of water, with a few swans and ducks floating peacefully on the surface, with an ever-present eye for handouts. The path leads us out of the park, and back into farmlands and nature.

"I'm heading to Ventosa today," I say when asked. "My comfortable limit is around 20 kilometers."

"Oh, we're hoping to make it to Najera," Karen says.

"Wow, that's another ten and a half kilometers." I'm impressed.

Previously I had toyed with the idea of trying to make it to Najera, and since we are enjoying each other's company I say, "you know, I might just join you." It's easier to do the longer distances when walking with others. However, I've also noticed I'm much more ambitious earlier in the day before my feet start hurting.

We round a bend in the road 13 kilometers into the walk and cross it. Before us lies a long, wide path, bordered on either side by endless fields of grapes starting their Spring budding. We stop for a moment under a brilliant blue sky to take pictures and wonder at the sight of a charming little town, Navarette, three kilometers in the distance.

By the time we get there, we are all in agreement that it's time for a café con leche. At the first bar, two pilgrims slouch tiredly in white plastic chairs at one of the outside tables. Their bare feet rest on top of boots, their backpacks lean against the wall. They are smiling and laughing as they recount their journey so far today. I begin to unsnap my pack, my body already relaxing at the thought of sitting down with a cup of coffee and fresh orange juice. But Eyleen, head bent over her guidebook has other ideas.

"This says there's a don't-miss bar in this town, but we have to go a wee bit further up to the top street to find it." She says in her soft Irish accent.

Oh-dear-God, really??? I look longingly at the chair I had laid claim to, and then up the steep road she is pointing at. But Eyleen is brooking no argument and I have to make a quick decision, as they are moving on. Oh, what the heck, they've been great company this morning. I re-snap my pack with a quiet sigh and up, up, up we climb to the highest point in the town (because there aren't enough ascents on the Camino already).

An older woman with a full apron tied around her calf length dress, leans over a fence and gestures in the direction we are headed. Speaking in rapid Spanish with a beaming smile and proud nod of her head, she is trying to convey something about a beautiful church up ahead. That much I can gather. Continuing on the one lane, cobblestone street, we soon come upon the towering, stone edifice with a bell that we assume to be the home of all this beauty. It is sandwiched between two ordinary looking buildings. I sense this is my place to say a prayer for Cooper today. My Irish friends decide to continue on and keep looking for the bar, which according to their guidebook should be coming up shortly. Eyleen is like a dog with a bone about this damn bar. "I'll meet you there in ten minutes," I say.

As I walk into the church, I murmur, "Holy Shit!" Before I catch myself and realize I've said it in my out-loud voice. Oops! I smirk and put my hand to my mouth, trying to push the disrespectful words back in. The art backdrop covering the wall behind the altar reaching up three stories, is all real gold plate, sparkling under the lights trained on it. It is breathtakingly gorgeous. While I expect to see this in the big Cathedrals, I'm continuously amazed at how much gold is in these little country churches. Now I know why the little old lady was so proud and excited to tell us about this.

I put a donation in the box, say my prayer for Cooper, then turn to leave. But out of the corner of my eye, I notice a statue of Mary and child in a little alcove with several pews in front of it. I stand before her and look into the kind eyes that the artist has captured, appearing to look directly at me. Her arm is protectively curled around her baby. I close my eyes and feel her vigilance. I ask that she continue safeguarding Cooper and say a prayer for myself also during the remainder of this journey. I drop an extra euro in the donation box on the way out.

I leave the church and keep climbing up the street looking for my friends. I suddenly realize that I have no idea where they've gone. I had been following them and had not thought to ask for directions. What had seemed like a small town is turning out to be a chutes and ladders game of little-cobblestoned side streets. I search for ten minutes but know after two that I am not going to find them. I can only smile, as I know then that I am not supposed to change my plans today. As much as I enjoyed their company, I surrender to the Camino's agenda. I know that this walk is so powerful, on levels more profound than I can imagine, and I just need to continue trusting the process.

A day later, I was told by another pilgrim who had seen me walking with the Irish lasses, that they had gone back to the church looking for me. It was nice to know they hadn't just successfully ditched me. It was also a confirmation that there was indeed some divine intervention as our ships passed in the night.

I'm so tired I need to find a bar to grab some lunch. The first one I come to is larger than I'm used to. The tables are neatly organized around a big room. As the only pilgrim in the place, I feel the eyes of the four old Spanish men, meeting for coffee, flicker on me curiously as I sit by myself. It seems that pilgrims mostly love gathering in small, intimate bars, where they have to jostle around each other, stepping over backpacks and trekking poles. Conspicuously missing in this room, is the boisterous camaraderie of people greeting like old friends, someone they met only a few days ago.

With no one to talk to, I don't linger over lunch. I'm back on track to keep my reservation in Ventosa and would rather arrive early and get horizontal in a bed for a while. I step onto the sidewalk wondering which way to go and notice a gaggle of French and German pilgrims noisily crossing the street. I follow them and am relieved within a few minutes to see a yellow arrow, hand painted on a building at eye level, guiding me in the right direction. I've spent enough time being mildly lost today, my feet are hurting, and I'm glad to be on the final stretch. I am quickly swallowed up in their group but relieved that they are studiously ignoring me, a ghost in the crowd. I was talking most of the morning, and it feels good to find some quiet time amidst their foreign chatter.

As we approach the edge of town, I hear the un-mistakeable clip-clop of a horse's shoes coming up from behind us on the opposite side of the street. I turn and blink. It looks like a Lipizzaner, a world-renowned, very-expensive breed of dressage horse, he is all white including a silvery mane and tail. The brand on his hip confirms he is special. He's high-stepping proudly, his rider smartly dressed in a black, long-tailed suit. Not the kind of horse or rider I would expect to see trotting down the street of a small, sleepy town in Spain. No one else in the group seems to appreciate how extraordinary this is. Indeed, they seem to not notice at all. I find it interesting that several people can be in the exact same circumstances and have a completely different experience.

The European posse eventually moves on ahead of me. I walk on my own for a while, before being joined by a retired pediatrician from England. She is an engaging woman and helps the remaining, uninteresting road miles drift by. She spots a large sign in Spanish next to a dirt road, that looks to me like another advertisement for lodging.

"Isn't that where you're headed tonight?" she says, pointing at the sign.

I didn't realize that while the Camino route used to go through Ventosa, it now bypasses it if you are not planning to stay or visit. Left to my own devices, I would have walked right past this turn-off. My angels have their work cut out for them today to keep me on track.

Something happens to my body when I know I'm nearing the end of my walk for the day. My legs suddenly feel like lead, my feet are in a lot of pain, and that last kilometer seems to take for-ever. This is compounded now by my inability to find my small hotel, in spite of the Lilliputian-sized village consisting of less than six streets. After I get three different directions from as many people and am ready to give up, I finally stumble upon it quite by accident. I realize that *Hotel,* might be considered a somewhat-grand name for this lovely, but unassuming, private residence.

I am greeted at the door by my hostess. She is a poised, middle-aged woman, her dark hair drawn into a neat bun, slender legs accentuated by a pencil skirt and high heels. Everything about her screams class. I catch her quick, head-to-toe glance summing me up.

Too tired to care if I pass inspection, I gingerly step into the immaculate, expensively adorned, entry area. I quickly remove my dusty boots and place them in the boot rack. She gives me a kind smile and asks me to follow her into the next room. There, an equally well-dressed man of the same age, I assume it's her husband, is sitting behind an antique wooden desk and accepts my payment.

The regal woman then takes me upstairs to a beautiful room dominated by a queen-size bed with matching end tables. The linens are so clean I'm afraid to touch them before my shower. I start to put my backpack on the Queen Anne chair in the corner, think better of it and place it on the floor. After a couple of weeks at spartan albergues, more reminiscent of barracks than a bedroom, this almost seems a little self-indulgent.

The hostess informs me that dinner tonight is salad, *their specialty,* (I could swoon with delight) and homemade paella. Things are looking up. Several hours later, after a marvelous shower in my en-suite bathroom and some feet-up time, I dress in the cleanest stuff I have and go down for my much-anticipated meal. I'm directed to a small living room with a couch and a couple of chairs, already occupied by a few other pilgrims. We politely greet, but the formal atmosphere leads only to quiet, stilted small talk.

From where we are sitting, we can see the lavish dining room, with a large table set with fine china and wine glasses, promising that dinner will be a civilized affair. There are eight of us, only one I haven't met before. We all wait politely to be called to dine. Considering that over 200,000 pilgrims walk the Camino each year, it's quite astonishing how frequently you run into the same people. Red and the Danishwoman duo are here. Red took a taxi for half the way today, her feet are unbearable, she doesn't look happy. I sit across from them at dinner, and we chat. But Red's boisterous sense of humor is subdued. Sadly, it seems, in spite of their great friendship, their walking time together has run its course. Red is suffering and clearly needs to stop, Danishwoman is definitely ready to move on. I will not see them again. Leaving this day on a high note, dinner was one of the best I had on the Camino. You can't beat home cooking.

Day 12: Azofra

After the guilty pleasure of a night in a sweet hotel room all by myself, I hobble off this morning down a quiet street, on very sore feet. No one appears to be stirring yet. The sky on the horizon ahead of me is momentarily the color of soft pink roses. I'm in awe of these magical few minutes of dawn, always too brief in their beauty and stillness. I quietly chant *E ala E,* the Hawaiian greeting to the morning sun.

The bar at the edge of town is open, I'm happy to give them my patronage for a café con leche and banana. I leave the village on another wide, gravel path. I am soon passed by a man about my height and age. His kind eyes speak of his gentle nature. He has the trim body of someone who walks all the time and totes a stick taller than he is. His full head of gray hair is a little too long to be sticking straight up from his head, but it does anyway, giving him a slightly startled look. Who am I to talk, my hair has been pulled back in a ponytail for two weeks now, and my odd assortment of clothes is quite another story.

Hans is from Germany, his English is impeccable, and he slows down so that we can chat for a while. His wife was unable to join him, so he came solo. After the usual introductory exchange, we both agree to some silence and separate walking, to fully appreciate the environment during the golden time of the morning. Hans walks faster than me but likes to stop and take pictures, so we play leapfrog, chat – silence – chat – silence – chat – silence.

In some ways, the walk is unremarkable this morning compared to other days. But Mother Nature never fails to please if you pay attention to the details. The serrated ridge of snow-covered mountains in the distance illuminated by the sun, makes a stunning backdrop to the endless tangle of grapevines. Today we begin our walk through the Rioja district, world famous for its wine. Hans attempts to capture natures perfection with his camera. He is down on his knees, agonizing over his shot for minutes at a time. This is in sharp contrast to me pointing and shooting with my iPhone barely stopping as I stroll past him.

As Hans and I approach Najera, we are walking together again and decide to stop for second breakfast. Then I go off in search of a Farmacia. My left foot is hurting, I'm not sure what's going on, my diagnostic skills have abandoned me. But I'm pretty sure ibuprofen will take care of whatever it is. I now get to add taking over-the-counter painkillers, to my list of; *things I don't normally do, but I am doing on the Camino.* I'm going to walk anyway, I don't feel the need to suffer if I can help it. The smallest dose they sell at this Farmacia is 600mg. I know I could easily do with 200-400mg. I ask if I can cut the pill in half. With a smile, the pharmacist helpfully tells me I can take two if I want to. I chuckle at yet another miscommunication, say a prayer for my liver and stomach and pop one right there. I feel almost immediate relief – I'm open to it being the placebo effect, I don't really care. But still, I'm grateful to be a modern-day pilgrim with access to drugs. I think my angels are having fun slashing one after another of my rules, for this trip and my life. My horizons have now expanded to include pharmaceuticals, even if they are the over-the-counter variety.

As I leave Najera, I run into Hans again, face tilted up, admiring what look like eagles soaring above the cliff faces behind the town. The Sedona-ish, red rock formations rising several stories high, are as spectacular as the birds. We continue to walk together and alone, joined by others in fleeting conversations. "Where are you from?" "Where did you start?" I notice that asking someone's name is often last on the list of questions, and in short exchanges seldom comes up. The Camino is like a mini-society, with its own structure and rules, you learn them by being there and paying attention.

At one point along the path, I'm very grateful to be walking with Hans, as we pass a truck parked randomly in the middle of nowhere. There are fields on either side for as far as the eye can see. Two young men lounge next to the truck. They could be farmers discussing their crops, but it doesn't feel that way to me. They "Buen Camino" us as we walk past, but I know I would have been a little nervous had I been on my own.

I quit at Azofra today. 18 kilometers is a relatively-short distance, but my foot needs a rest. Hans decides to stop too and joins me at the municipal albergue. *Munies* are often large, crowded, noisy and cheap. This one is large and cheap and will later become crowded. But at only seven euros, with the added luxury of small, semi-private rooms containing only two beds and no bunks, it's a screaming bargain.

I empty everything out on my assigned bed. I have now surrendered to this as just another part of my ritual, instead of viewing it as an inconvenience because I can't find anything. I grab some clean clothes and head for the shared washroom facilities. The bathroom is small and feels more intimate than most. Standing at the sink in his red, plaid boxer shorts is Hans, brushing his teeth. Like strangers on a crowded subway in New York, we barely acknowledge each other to give a modicum of privacy.

But still, I need to squeeze by him to get to the three showers that have no locks on them. I enter one and undress, aware that Hans is standing on the other side of the door, a mere two feet away. There is nowhere to hang anything. So, I remove my dirty clothes, then fold and pile them precariously on my baggie of toiletries, that is sitting on the wet floor in the corner. I don't know why I care, I'm going to launder them later anyway. I sling my clean outfit over the top of the door, my *Do Not Enter* sign. To turn on the water, I have to push a button. An unimpressive stream comes from a showerhead that they have somehow managed to mount almost flush to the wall. If I wanted to wash my hair today, I would need to lay my cheek against the surface of the shower stall and rotate my head around to get it all wet – I think not! It's enough that I have to get up-close-and-personal with the bathroom wall to clean my body. I also have to keep pushing the button every 30 seconds. I appreciate their water conservation efforts, but it wasn't the most satisfying shower – at least the water was hot.

It was a short day, and the ibuprofen is still working, so I have some energy to walk around the tiny village in search of a church. It is closed, but I sit quietly for a few minutes, say my prayer for Cooper, then go to look for a bar for a snack. I sit outside at one of the two plastic, green tables in a matching chair, with a fresh orange juice, and soak up some late afternoon rays. I marvel at the luxury to relax with no electronic devices and do absolutely nothing. With a cool breeze at my back, I feel such gratitude for this prolonged period of stress-free living.

I return to the albergue, hand wash a few items of clothes and hang them to dry in the sunlit courtyard large enough to hold a party. In the middle is a small, square foot-soaking pool, tauntingly empty of water. Pilgrims are scattered everywhere, like contented cats, quietly enjoying the warmth of the sun. I find two plastic chairs and sit on one, as I put my feet up on the other and take out my phone. Good, I have WiFi today, I'll catch up on my blog.

Hans finds me outside and invites me to join him for dinner with two other German men he has met. As we enter the bar, a man with a broad smile waves us all over to his table. He has shoulder length, brown, unruly hair – handsome in a dark and broody kind of way. How is it that I still find men like this, men that have *trouble* written all over them, attractive? I have noticed him several times on the trail in the past couple of days, occasionally looking at me a beat too long. He sits with a younger man who turns out to be his 20-year-old son. They are also German. I dub this *my night with the Germans*.

The pilgrim meal tonight is forgettable, but we hardly notice with the free-flowing wine and lively conversation. They are kind to all speak mostly in English so that I can follow. I put my hand over my glass as Mr. Dark and Broody, sitting across from me, reaches over to try and refill it for the third time while giving me *the look*. As he openly flirts with me, more out of habit than actual interest I suspect, I feel a hand on my shoulder. Hans has casually put a possessive arm around me. What is going on here? First the Marlboro man and now this! I haven't had this kind of attention from men in 13 years. I'm so out of practice. A part of me would like to participate in this flirting ritual just for fun. But I'm as socially clumsy as a kindergartener on the first day of school. I just don't remember how. I am also acutely aware that my unwashed hair is pulled back tightly in a ponytail and I'm not sure of the last time I plucked the hairs on my chin – *did I even pack my tweezers?* I'm not prepared in any way, shape or form for a night of flirting. Besides, two glasses of wine are one over my limit, and I'm feeling decidedly light-headed. On this note I think it is time to retire.

Day 13: Grañon

Following a restless night, I make the decision to leave by myself this morning. The overcast sky fits my introspective mood. I welcome my alone time during the magic morning hours on the trail this morning, to re-focus. As much as I enjoyed Hans's company yesterday, he's a married man. I didn't come on this trip looking for romance, and I feel he might be getting a little too attached.

As I see Hans heading towards the bar in the village for breakfast, I decide this will be a good way to casually put some distance between us. And I make the big mistake of not stopping for coffee. Leaving Azofra there is a small statue of Mary and Child behind glass built into a square brick monument. I stop to acknowledge and say a quick prayer for Cooper.

The first town Cirueña, is a peaceful 10.5 kilometer walk of blessed nothingness. These vast distances of paths and nature are a rare luxury in today's crowded world. I am so overdue for my first cup of coffee, but Cirueña appears to be a ghost town. Oh, this is the place I've read about. A golf course and row upon row of abandoned townhouses, built years ago before the bottom fell out of the real estate market.

I pass an unappealing bar at the golf course thinking there must surely be another. But I learn you can't really tell from the map what is in store for you in each village. Before I know it, I'm leaving sans coffee or food. In hindsight, I wonder why I didn't just retrace my steps a little for breakfast when I was indeed hungry, and seriously in need of a café con leche. Why did my body just keep moving?

The CCCC (Cosmic Coincidence Control Center) was never busier than when I walked the Camino. When I think of the critical timing necessary to put me in the exact right place and time to meet the important people I needed to on this journey, it's nothing short of miraculous – and this was one of those times.

At least the sun has now made an appearance. For some reason, the Hawaiian forgiveness prayer, ho'oponopono comes to mind, so I recite it. Not for anything in particular, just covering my bases. I leave Cirueña on a wide trail between farmer's fields leading to the horizon. One golden rectangle of yellow rapeseed in the distance breaks up the monotony, if you can call it that, of emerald green, young wheat surrounding it. A man over 6 feet tall and a few years younger than me, with short, thinning hair and glasses, walks up behind me. His openness and smile embrace me like the morning sun.

Following Camino protocol, he begins with, "Where did you start?" And moves on to, "Where are you from?"

"Wait – what – Kona? On the Big Island?" He says with a grin from ear to ear.

"No way!" I say as it dawns on me why his comfortable, relaxed energy feels so familiar – there is actually something else I recognize, I can't quite put my finger on. It will be three weeks before I realize he bears an uncanny resemblance to an old boyfriend. Not just physically, but also his personality and mannerisms

Kevin lives, as it turns out, 20 minutes away from me and we know some of the same people, and I've been into his workplace several times. How is it possible we haven't met in Hawaii before, but like two needles in a haystack, find each other in the middle of rural Spain? We walk on for a kilometer reminiscing about back-home, before his long legs start itching to go faster.

I arrive in Santo Domingo, a place I had planned to stay, at 11:00 a.m. I am greeted by a small marching band walking towards me on the narrow cobblestone street playing lively, Spanish music, with more volume than tone. Well bah humbug! I'm not a fan of loud music or big parties. So, I'm even less impressed when I discover this place is in the middle of a three-day festival. Which means lots of big parties and loud music.

It's a good-sized town, so what are the odds that I will walk into a bar for my much-awaited café con leche, to find Kevin. He's sitting by himself, his cheerful demeanor is subdued. A chronic injury on one of his legs from running marathons has flared up, leaving his whole calf red and swollen. I join him with my coffee, and when asked for my opinion, concur that he should seek professional care. He is not happy about this but had already come to the same conclusion, and leaves before I finish to find a doctor.

As we say goodbye for the second time today, I assume we will not meet again, such is the transient nature of the Camino. I leave shortly after, determined to make it to the next town in spite of my painful feet. It's still so early, and I need to get away from all this noise.

Glorious sunshine and a slight breeze make for perfect walking weather. The decision to keep going is a no-brainer. As I turn down the narrow street to go in search of my next yellow arrow, coming towards me is a procession, on foot, in cheerful disarray. A dozen or more women of three generations, all wear lacey, floofy dresses and hair adornments, and are laughing and talking loudly. They surround a teenage girl who is either a bride or having a religious rite of passage that requires a white dress, maybe communion. She seems young, so I lean towards the latter. I look to my left where they are headed and see that the party has started without them. A large group is gathered in the square, musicians playing and head bobbing as they weave through the rollicking crowd. I laugh out loud. Partly I'm sucked in by the joyful abandon surrounding me. But really? This discordant music again? It's clear that the entertainers are having fun, so maybe that's their end goal. Am I the only one Scrooge enough to notice that either their instruments aren't tuned up, or they don't practice enough?

As I go around a corner, the Cathedral of Santo Domingo de la Calzada looms on my right. The name niggles in the back of my mind. While I read copious books on the Camino, for some reason, much of its history just refuses to stick. Bits and pieces rise to the surface occasionally, like the flotsam from the early stages of homemade chicken soup. This Cathedral is famous for something, but it eludes me for now. I duck in to escape the mayhem outside, and in the hope that maybe this church's importance will reveal itself to me.

This Cathedral isn't messing around, they're not going to rely on the generosity of pilgrims, some of whom think *donation* means *free*. To see anything in the church, I need to pay three euros to the bored young man behind the desk and leave my backpack in a little room with half a dozen others. I hesitate only briefly. I don't begrudge them the three euros, but I'm a little uncomfortable parting with my pack full of all my current worldly possessions. I glance over my shoulder as I abandon it in the dark, lonely room. I'm coming to understand why homeless people are so protective of the few belongings they have.

Well, that was three euros well spent. What a beautiful Cathedral, the gold and priceless relics are breathtaking. I sit in one little alcove under the kind, beneficent gaze of a statue of Mary. I pray that she projects her maternal protection all the way to the East coast to envelop Cooper, and comfort him during his arduous journey with chemotherapy.

As I'm leaving town, a word pops into my head, *Chickens!*
What?
The chicken myth ...

Oooh, that's what the Cathedral is famous for – darn. I even heard they keep a couple of live chickens on display in a crate somewhere, to commemorate an ancient legend. I'll spare you the story, you can find it in most other books about the Camino. Somehow in my three-euro tour, I missed that. All for the best. I'm sure I would have obsessed over the short straw those poor chickens drew in this life.

I'm back on the road and it's not even 12 o'clock yet. My feet feel remarkably good, so I walk with purpose for an hour. The terrain is kind, my poles tap-tap a soothing rhythm. The sunshine that was nurturing is fast becoming just plain hot. Up ahead is an eight-foot-tall cross. Beside it, are two wrought iron benches. It's definitely time for a break. And sitting on one of the seats is a pilgrim I met yesterday, that seals the deal.

The notable feature of this classic looking, East Indian man, with short, dark hair, is that the hiking boots he is wearing are duct taped together, I'm talking seriously duct taped. While a brand-new pair hangs from his backpack.

I decide to get right to the point, "So what's with the duct taped boots?"

He smiles proudly, "This is their third Camino. I brought new ones with me in case I need to switch them." He points to the ones dangling from his pack.

I laugh because I can't help myself, even though it might be seen as rude, "What are you waiting for?"

"Well, if I can make it to the end of this Camino with my old boots, I'm going to donate the new ones to a charity." He says beaming proudly.

I nod as if I understand, although his whole scheme is a little lost on me. There are so many different ways to donate to charities – a pair of hiking boots seems pretty random. But I can see that for some reason it's important for him to carry the added weight he's hoping not to need, so that he can eventually give them away. I have noticed, even for myself, that when traveling with so few possessions, there is a tendency to form unusual attachments to the ones you have.

After several kilometers of wondering why I didn't just stay in Santo Domingo, Grañon finally appears in the distance. It is of course at the top of a rise. Sigh! My feet are complaining loudly, so I concentrate on my poles click-clacking a steady rhythm to carry me there. What an accomplishment, this was my furthest day yet – 25.5 kilometers – and notwithstanding my feet, I am less tired than on previous days.

I have a reservation at a private albergue. But when I find out it is a further 2 km walk, my body begs me to see what else is available. Grañon appears to consist of a single-lane cobblestone street. It is lined with a church on one side, a tiny store with a couple of tables on the other, and further on a handful of residences and other small businesses. I know there will be few choices.

At the entrance to the village is a small gathering area, with three old local men passing the afternoon. I approach one of them, and after greeting him with "Hola, Buenas Dias." I ask "Albergue?" In the rapid barrage of Spanish that comes at me, I recognize the word *"donativo,"* and that is enough. I know that these are albergues with no set price, the room and often dinner is by donation. I had not really planned to stay at any of the donativo albergues, sprinkled liberally along the Camino route. They tend to be very bare bones and cater to people doing this walk on a restricted budget. But I am so thrilled at the prospect of a bed in spitting distance, that I decide to add this to my list of experiences. The old man's hand gesture pointing down the only street, makes it easy to find.

I recognize the American hospitalero's shaved head with a tiny ponytail in the back, I think this means he is a Krishna devotee. It explains his dedication to service and life of near poverty, in this funky old building. He later tells us, over dinner, he has walked the Camino many times, the most recent he did in 23 days and will be his last.

I follow him up the sloping, uneven stairs to the third floor, where he gives me a choice of the three beds left. I choose the single bed in the long hallway, with three other male inhabitants. The other choice was a tiny bunk, in an even tinier room with no windows and a distressed, naked lightbulb hanging crookedly from the ceiling. As he leaves without a word, I take a deep breath. I find gratitude for the window next to my bed letting in natural light and use that to search carefully for signs of bedbugs. Of all the places I've stayed so far, this would seem the most likely to have them. I am in for one pleasant surprise though. As this is an old converted house, there is only a single bathroom, with a lock on the door, giving luxurious privacy.

Nothing about my bed for the night makes me want to hang out here and rest like I usually do. So, after my shower, I leave to explore the village in the gorgeous afternoon sun. I have only passed a few buildings, when I recognize the smiling face of Hans walking towards me. There are no recriminations, we all respect each-others need to *walk their own Camino*. So, we don't take it personally when pilgrims come and go from our lives. Conversely, I suspect there are no accidents on this powerful pilgrimage, so I have come to honor all chance meetings. He proudly entreats me to come and see his room in the loft of the church, another donativo albergue. He tries to talk me into switching residencies after I tell him of my sketchy accommodations. After ascending the narrow staircase, I see why he is so excited by his find. An inviting communal area with a fireplace, dining table and adjoining kitchen, where they will all share a jointly prepared meal this evening. The stairs up to the loft reveal a large, open room, with mats and sleeping bags all crowded together on the hardwood floor. I'm sorely tempted, this place has a very homey appeal to it. But once again, I note a slight magnetism between us and think it best to keep some space.

Downstairs the Church calls me. While it is missing some of the grandeur of the Cathedral in Santo Domingo, a subtle energy embraces me as I walk through the door. I know this is another place I need to pray for Cooper today. The gold behind the altar appears to be standard in this part of the world, regardless of the size of the Church or congregation. I have been wanting to attend a Pilgrims Mass and discover there is one this evening at 7 o'clock. This tiny village has a palpable magic, and deep spiritual energy. I intend to enjoy as much of it as I can.

Across the street, I greet Yoko, the Dutch woman from a few days ago, who is also staying at my albergue. She is outside having lunch with another European woman whose dismissive demeanor tells me I'm not welcome. So, after I order my food, I decide to sit inside with another pilgrim I met the day before. John is a lawyer, who knew? His story is like so many, of searching for more meaning in life than his job. Sitting at the next table is our hospitalero, eating pizza and playing chess. I wonder why he's having pizza when dinner will be served in a little while – I will find out soon enough.

As 7 p.m. nears I enter the church and see that it is filling up fast. I don't like to sit in the back and notice that the front row is empty. With a raise of my eyebrows and a shoulder shrug, I think, *I don't mind sitting in the front row*, and I take a seat. After five minutes, I realize I am still the only one there. I turn and see Hans entering the church and beckon him over to sit with me. With a horrified look, he shakes his head, "no." *Hmm, does he know something I don't?* I now notice that the three, straight-faced, old local women in the pew behind me, are staring at me like a cat watching a bird. I ask with the aid of hand signals, "am I am supposed to be sitting here?" Given the quick response, I realize the unspoken leader has been waiting for her opening to let me know I have broken some unwritten law. With a shake of her head and a stern, "No!" I get the message. And although I clearly don't understand the language, she continues to scold me in rapid Spanish as, just in the nick of time, I quickly squeeze into the row behind them, beside Yoko. I never do figure out why I wasn't supposed to sit there.

I gasp audibly as the priest walks in from the side, and simultaneously the altar and backdrop are lit from all angles. Talk about making a grand entrance. The glittering sparkle as the light hits the gold, going all the way to the high ceiling, would do a Disney castle proud. My embarrassment from my earlier faux pas is short-lived, as I enjoy this visual banquet, and a thankfully short mass and pilgrim blessing.

I return to the albergue just as they are preparing to serve the evening meal. I am the last of 14 people to be seated. The chill in the room requires that we all keep our jackets on to eat. Somebody passes me a plate with the nod to something green, four shreds of iceberg lettuce and a tiny sliver of tomato, no dressing. This is followed by roasted potatoes. I'm guessing they're the highlight of the meal, as they are limited to three chunks per person. Slices of French bread are passed around. Then a giant pot is placed in the middle of the table. It is filled with white rice, some peas and carrots for color and a few tiny pieces of ham (kind-of-like a poor-mans' paella). I look over my shoulder to see if the rest of dinner is on its way, only to realize this is it; bread, potatoes and rice, a white, starchy supper. A donativo meal is paid for by the take from yesterday's pilgrims. I make sure to leave a generous donation when I leave in the morning, in the hope that tomorrows recipients might receive a more exciting dinner. I am given an opportunity to reflect on my expectations of variety and abundance in today's world and appreciate the relative affluence of the majority of Americans.

Day 14: Tosantos

At breakfast, I snap off a chunk of yesterday's, now stale bread, and make a mental *note to self*, to avoid the donativo albergues going forward. I appreciate that these places make the Camino affordable for people on a limited budget, so I will leave the beds for them. That said, the deep spiritual energy of the albergue is strong and sweet. For me, this was an opportunity to experience a little of what it might have been like hundreds of years ago for pilgrims.

The sky is threatening as I take off at 7:20 this morning. Ten minutes later, the line of pilgrims spaced out ahead of me all stop to don wet weather gear, as the promised rain begins. Today's path parallels a busy highway. I notice road walking sucks my energy, it doesn't give back to you like Mother Nature does.

The welcome mat for Belorado is a large bar whose driveway, lined with colorful flags from many nations, beckons me in. The sun has come out just in time to illuminate it and complete the effect. I notice Yoko sitting by herself at a table, in an otherwise deserted room. I take my coffee and tortilla and ask if I can sit with her. Barely looking at me, she nods her head *yes*. My attempts to extend the hand of friendship to her, have so far been unsuccessful. I seem to run into her frequently, so I keep trying, but the one-sided conversations are painful. It's hard to know if it's the language barrier or something more. I never do figure out if this woman is just painfully shy, or if she has her own reasons for remaining solo on this journey. I sense an air of melancholy around her. I was thinking of overnighting here, but it's only 11:18 and other than this bar, the town doesn't call me at all. It seems silly to stay, so I continue on to the next village, Tosantos.

My feet are really starting to hurt. So, I'm relieved when finally, the path veers off the highway and once again runs between farmer's fields, with at least some visual mana for the soul. I turn a corner to what I have started to call the *Hallelujah moment*, my first glance of the Town/Village I will stop for the night. This does not mean that the place is close, it can still be anywhere from 1-5 kilometers away, but at least it is in sight. Thank God this time it's not far. As I enter Tosantos, I know immediately I have made the right choice.

Tonight, I stay at a brand new albergue, the polar opposite of last night's experience. Thankfully, they have a bottom bunk left in a room with nine sets of beds. The adjoining bar is a bonus, I won't have to go looking for dinner tonight. The cheerful young lady bustling about readying the place for the evening, proudly explains in a spare moment, that they only recently finished building this family-run business. She and her brother, like so many other albergue owners, will labor 24/7 for 9-10 months of the year. They will only take a short time off when they close for the winter. These industrious hospitaleros all along the route, give work-ethic a whole new meaning. For minimal sums of money by American standards, they turn over numerous beds and prepare meals every single night for the hordes of pilgrims.

I wash clothes in the bathroom sink and hang them out to dry in the bright sunshine. Following my shower, I grab a glass of wine at the bar and sit outside on the lawn in the courtyard to write my blog. Traveling alone is forcing me to dig deep and find my social side. When two women who look a decade younger than me, take chairs nearby, I introduce myself and ask if I can join them. I seem to be meeting lots of pilgrims from Canada, eh?

They give me, what I consider, a lukewarm reception. But I decide it's cultural when they invite me to have dinner with them. Living in Hawaii, I'm used to open friendliness even from strangers, so I have to adjust my expectations a little on this trip. We unanimously decide to also ask an older Japanese woman, who is sitting by herself at the next table, to sit with us. She is tiny and shy, curled in on herself and wears a kerchief over gray hair. She is at first quiet, but suddenly with a mischievous smile, in broken English says, "Dis trip, um, my mother – surprise." We understand this to mean; her mother doesn't know she's doing this as she would disapprove.

She then adds, "my family – very worry – but I like do this long time." It's too bad there is such a communication barrier, we are only able to get the basics and I suspect there's a great story here. She seems quickly exhausted by the effort to express herself in a language she barely knows. After dinner, she goes back to sleeping in her bed. Lost and lonely, she is separated from most pilgrims by culture and an inability to be understood. To think I was afraid to do this trip on my own, this woman's courage puts me to shame.

I have never asked my body to do this much physical activity day after day. I'm so grateful that it's holding up. My feet hurt at the end of the day but taking 600 mgs of ibuprofen in the morning and at night is giving me hours of relief. I have surrendered to the miracles of western medicine for now. Since I don't really know what's going on, there's not much else I can do.

Day 15: Ages

My Canadian buddies and I agreed last night to walk together this morning. But they are ready before me because of my complicated feet taping regimen, so I tell them to go on ahead. When people make this journey with friends or family, they're not always open to others traveling with them. In my observation, solo pilgrims often find other *solos* with compatibility in temperament and walking speed. One of the unspoken rules on the Camino is to have no attachment to any relationship, no hard feelings when it's time to move on. This can be after ten minutes or ten days. Rene, a friend I have yet to encounter, calls this the *Camino reset*. Her policy was to walk with someone no more than three days, so that she could take maximum advantage of this opportunity to meet new people.

So, even though my Canadian friends have asked me to make bed reservations for them at the same place I'm staying tonight, I want to be sure I'm welcome to join them walking today, and they're not just being polite. I decide to let the Universe figure it out, and if I meet them on the trail, I will take that as a positive. On a selfish note, I also enjoy having the golden morning hours to myself, even though it immediately starts to drizzle and continues most of the day.

The walk to Villafranca Montes de Oca is peaceful. For a while, I am behind a Korean woman traveling with her daughter, who looks to be about ten years old. I run into them frequently for the first few weeks, but because of the language barrier and their unique little *bubble,* I never end up chatting with them. In fact, I never see them talking to anyone but each other. I am in awe of this beautiful mother/daughter bonding experience.

An interesting stone building in a wheat field gets my attention. There are so many strange, ancient structures in the middle of nowhere, pondering their history and making up stories about them helps to pass the time. I stop at the first bar for coffee, having learned my lesson about that, and run into my Canadian friends. They appear glad to see me, the energy feels good, so we begin the long, uphill path out of Villafranca together.

The initial two-kilometer climb is on a steep country lane lined by trees and fields. It quickly turns into a 13-kilometer walk, through a magical pine forest, with a few ascents and descents. We are far from the highway and are once again serenaded by the birds, as we dodge from one puddle to another in the drizzly rain. The path in many places is wide enough to land a small private plane, but the mud is relentless and impossible to avoid. The addition of a couple of pounds of wet, sticky sludge on my boots adds to today's challenges. The pain in my feet has kicked up a notch, and I can feel my gait is different in an attempt to distribute my weight differently. I recognize the same rolling motion in others and realize there are many of us suffering in silence.

I am surprised at how small San Juan de Ortega is when we finally arrive. It is one of the recommended destination villages on the Camino, one of Brierly's *stages*. But there seems to be little more than a monastery, with a large albergue and bar.

All three of us found the trail challenging today, primarily the previous 13 kilometers with no breaks. With the rain and mud, we couldn't even stop for a rest on the side of the road, and there were no food stops. So, we are relieved to sit and have a cup of tea and some freakin' amazing, Spanish chocolate, before tackling the remaining five kilometers to our destination for this evening.

The sun has finally come out, and the path ahead is just plain gorgeous. A combination of fields and light woods make this final haul on my wretched feet do-able. Before we know it, we have our *Hallelujah moment*. We walk gratefully into the picture-postcard village, to a clean, comfortable, well-run albergue, that is already turning pilgrims away. Thank goodness we made a reservation.

After a shower and a short rest, this tiny, ancient hamlet calls me to explore. I slowly stumble up the narrow street and turn a corner. There is a small, medieval church bathed in sunshine, with the now familiar sight of storks nesting in the tower, one is standing watch over her eggs/babies. I have discovered that storks love to build their huge nests in the bell towers of churches all across Spain. If the day is quiet, as you walk by, you can hear the curious clacking noise they make.

In spite of the minuscule size of this village, there is still gold beautifying the backdrop of the altar. I choose the pew in front of the statue of Mary and close my eyes. I suddenly feel a little light-headed. A vision of monks in brown robes and common people from centuries ago pops into my head. They are walking down the center aisle. I open one of my eyes a slit, then smile at my silliness, thinking I might actually see them. I wonder to myself if this place is haunted by friendly spirits. I say my prayer for Cooper, then sit for a while basking in the subtle, healing energy that I sense.

I leave the church under a sky that is now sapphire blue, my eyes alight on a little bench up against a building. I cannot resist the invitation to lay down for a while and nap. It is just the perfect cool temperature to be able to appreciate the warmth of the sun for half an hour.

I open my eyes to see a pilgrim setting up a tent in a grassy area across the way. I do not envy him. The walking every day is hard work, and I'm thrilled to have a bed and a meal waiting for me at the end of it, not to mention the camaraderie of other pilgrims.

The albergue tonight includes dinner in its cost, so after another short rest, I go downstairs to the claustrophobically small dining room. There are ten of us crowded around a table meant for six, we are, quite literally, touching elbows. The meal is excellent, and due to the intimacy of the seating arrangement we are soon laughing and telling stories like old friends. I sit across from Gudrun, a young, waif-like German woman I met a few days ago. Short, reddish-blonde, curly hair surrounds her beautiful face. Unfortunately, she is sick and will be taxiing ahead for a rest tomorrow. We make a strong connection in spite of the language barrier and the party atmosphere surrounding us.

Day 16 - Burgos

Today I have perfected my gangsta walk – step-limp, step-limp. The cause of this pain baffles me, it switches from one foot to the other, sometimes even going up into my ankles. It defies my diagnostic abilities, the severity of it leads me to believe it may be a hairline stress fracture – but on both sides? When I return home, my Naturopath diagnoses inflamed nerves, from overuse. But for now, I am in the dark and determined to not let the pain get the better of me. I override common sense (which if I were listening would tell me to rest), pop some ibuprofen and tough it out, relying heavily on my trekking poles.

I met Charlotte, an Australian woman, on my online journey to get here. The internet dissolved the thousands of miles between us, we became friends and considered walking together. But all the stars didn't line up. She decided to go with one of the eco-tourism companies that plan the whole trip for you. Her itinerary told her how many kilometers she had to do each day, when her rest days were, and which private albergue or hotel she would be spending the night. They would even send her bag on ahead for her every morning. They would do everything but walk it with her. Charlotte is my age, and all the uncertainties that are part and parcel of a journey like this, some might even say are what makes this trip special, felt overwhelming for her. It gave her a feeling of security and I suspect a certain sense of control to do it this way.

Some who consider themselves *real* pilgrims judge this method. I just weighed the pros and cons and decided it wasn't for me. I felt that the strong infrastructure that has developed over the years along the Camino, creates a safe container for the average person to have this magnificent experience. It was as much security as I needed. Having lodging set up ahead of time wasn't going to quell my biggest fear of being by myself in a foreign country in the middle of nowhere. Charlotte would still have to do that along with the rest of us.

Charlotte started two days after me, we stayed in touch by text and figured out that today our paths will cross. She and Harriett, a 75-year-old woman she met here using the same travel company, will traverse the five kilometers from San Juan this morning to meet with me, and we will all walk into Burgos together. My Canadian friends and I do a *Camino reset*. We exchange phone numbers and bid each other farewell. Although I make plans to call one of them in a week after the other leaves, it never happens.

With my backpack and poles on the chair next to me, I set my café con leche, fresh orange juice and a banana on the wobbly table outside the little bar. The table lurches and slops a small puddle of coffee into the saucer. But it doesn't even register on the annoyance meter, as I turn my face up to the warmth of the sun and breathe in a silent *good morning.*

I finish my breakfast just as Charlotte, who I recognize immediately from photos, turns the corner. She strides confidently towards me, wearing a grin that makes her eyes sparkle. Her sensibly short, gray and white hair gleams like a halo. We throw our arms around each other in a squealing, girly greeting. Harriett, a stern-faced Aussie, looks on unimpressed with a detached smile that says, *are we done? Can we go?*

I take the last sip of coffee, quickly slip into my backpack, grab my poles and start to hobble in my rolling gait after them. Two things become immediately apparent: One, that they both walk much faster than me, and two, that Harriett and I have no connection. Although she's older than both of us by 12 years, Harriett takes the lead, setting an alarming pace. I know there will be hell to pay for me not going at my comfortable speed. But Charlotte and I have anticipated this meeting for months, and my comfort level does not appear to be on anyone's radar, so I suck it up and match their rhythm.

I've given up my quiet, golden hour this morning to chat, so we almost miss the importance of the over-sized statue of a prehistoric man as we approach the small hamlet of Atapuerca. Three kilometers off the trail is an ongoing archeological dig of the earliest human remains found in Europe, dating back some 1.2 million years. Even they knew this was a special place.

We turn left out of the village and are presented with our only ascent for the day. A steep hill with no discernible top for now, with a rough trail going through scrubby bushes and alongside a farmer's field. My body is getting so strong in spite of my feet, which weirdly prefer going uphill. I've learned to set a steady rhythm and enter a light meditative state to scale these numerous inclines.

Nearing the top, Charlotte says, "Do you hear that?"

"Yeah, wow, what is it?"

The musical tinkling of cowbells and the low groaning calls of many animals is getting louder as we approach the crest of the hill. On our right, a farmer is herding hundreds of sheep with large, clanging bells around their necks towards the trail in front of us. After a quick appreciation stop, we realize we can't dawdle, or we might be stuck behind them for a while. Farmers here are business as usual, and let pilgrims take care of themselves. The blessed descent that will take us into civilization gives me a not-quite *Hallelujah moment*. We get a glimpse of Burgos far, far in the distance.

Our fearless leader has finally stopped to rest for a few minutes, allowing us to catch up. "So how did you end up coming on this trip, Harriett?" I ask in an attempt to be friendly.

"I like to walk," she says with a smile.

Okaaay, is this Aussie humor, or is she just a woman of few words.

"Do you have a family?" – let's try again.

"Yes, a daughter."

"There's a more scenic route we can take to get into Burgos to bypass the industrial area," says Charlotte, thankfully interrupting our awkward discourse. She looks down at her guide. "It's a couple of kilometers longer, but the turn off should be right ahead."

"Hmmm" I whine, "I don't know if I can walk any further than I have to today, my feet really hurt."

"It's right by the airport," she says still looking at her book "it shouldn't be too much further." – I realize she wasn't asking.

Harriett says nothing.

These two have an interesting dynamic that seems to be working well for them. Charlotte is a take-charge kind of gal, and while I sense a solid core in Harriett also, I suspect she's what my mother would have called *still waters run deep*. In other words, I have no idea what's going on in her head, but she is content to let Charlotte lead the way. I feel like a bit of an intruder in the bond they've developed over the past few weeks. Once again, I'm on the outside looking in. I don't know quite what to think about this, I just note it appears to be a re-occurring theme. While I believe most people would say I'm socially capable, I often lean on silence to mask my awkwardness and fear of saying the wrong thing.

"Well, I think we missed the turnoff for the alternate route," announces Charlotte, as we arrive at a busy highway. I'm relieved. Harriett's expression is unreadable. We find out later that due to yesterday's rain, the alternate route was four kilometers of thick, boot-sucking mud today. Score one for my angels. The missed turnoff is soon forgotten, as our immediate dilemma becomes crossing a busy, four-lane highway to get to a bar in our sights.

As I bring my long awaited second breakfast to an outdoor table, I look longingly at pilgrims boarding a bus to take them into Burgos. Word is out that there are few beds available tonight, so the rush is on to secure one. I am once again grateful that I reserved ahead. The only room I could find though was at the upscale Abba Hotel, so this will be a splurge.

We continue on a sidewalk that runs parallel with a four-lane highway for ten-and-a-half kilometers, through the industrial area. It is boring and hard on the feet. Chain link fences delineate parking lots and contain people in their *work prisons*. As my poles tap, tap on the sidewalk, I peer through the fence and reflect on my decision to exit a life like this 22 years ago.

As a fairly grounded Taurus (although my friend Rene thinks this is an oxymoron, using the word *grounded* in the same sentence that I announce my horoscope – this produced a belly-laugh session while walking one day), *I wasn't prone to making major life decisions based on a*

psychic reading. But I did just that when I left my job of 14 years, as a CAD
designer at Kodak, in Rochester, NY.

On a beautiful, summer day, I got a pivotal, clairvoyant reading at Lilydale,
a well-known Spiritualist community just south of Buffalo, New York. I took it
to heart when towards the end of the session, I was told, "I am seeing you on a
sandy beach with palm trees – you will have an opportunity to go on this
vacation— make sure that you do, it will change your life."

On my way home from the reading, I felt like she meant for me to go visit
my friend Katherine who had moved to Hawaii 8 years earlier. The ensuing trip
six months later did indeed, change my life. Fifteen months after returning from
the vacation, I had sold, given away or thrown away everything I owned, had quit
my job and moved to Hawaii to live.

A sharp pain through my right ankle brings my thoughts back to
the present and a fleeting memory of the small group of pilgrims at
the last coffee stop, boarding the bus to Burgos. I wouldn't have
missed anything bypassing this part of the journey (an option
suggested in the *Brierly* guidebook) and could have saved my feet
continued injury had I joined them. But I have company today, that
helps a lot, although Charlotte and I long ago exhausted conversation
topics, so the three of us now walk mostly in silence.

Just as it feels like this section will never end, we cross an
intersection to find ourselves in the new part of the city. All cities in
Spain seem to have their heart in a centuries-old sector, with the new,
modern portion encircling it, then the industrial section outlying that.
A car honks, narrowly missing a pedestrian crossing the street,
nobody seems to notice. We are shoulder-to-shoulder with people
weaving in and out of each other going in both directions. No "Buen
Camino's" here, just people on their way to somewhere, studiously
ignoring most of their surroundings. Stores line the narrow sidewalk,
signage in every color and size adorn the windows and *OPEN* signs
flash. The air is scented with exhaust fumes, occasionally
overpowered by a passerby's strong perfume and the white noise of
traffic as a backdrop. The assault on our senses heralds our entrance
into what could be a big city anywhere on the planet.

My search for an ATM helps to keep me grounded in my boots.
I cling closely to Charlotte, sure that if we become separated, I will

soon be hopelessly lost. I've gotten myself this far just fine. When I'm in nature, I'm in my element, relaxed, observant and tuned in to my intuition. But in a big city, it's like something in my brain turns off. So, I'm more than happy to let someone else lead in this foreign-to-me environment. I once again feel the presence of my angels protecting me and sending me all the resources I need for this journey. Today that would be Charlotte.

I usually just follow yellow arrows, but Charlotte has put herself in charge of the map. She's reading all the fine print in her *Brierlys*, making decisions on alternate routes, etc. I notice she's not asking for input or opinions. Fortunately, I don't really care, my only criteria is *what's the shortest route?* Besides, following the yellow arrows becomes a little more challenging in a big city, with so many other things vying for your attention. I'm grateful she's willing to take this on.

If it's possible, the old town is even more saturated with people. But the energy is different. It feels like we just entered the Party Zone. Then I remember, *right, it's fiesta weekend (another one)*. The now familiar picturesque buildings and narrow, cobblestone streets are throbbing with people. Suddenly, in the distance, several spires rise above the rooftops. There's no doubt this building dominating the skyline, is the famous Burgos Cathedral.

In the midst of this celebratory energy, however, I'm disturbed by the plight of so many of my fellow pilgrims. They wander around aimless and dejected, not knowing what to do. Fiesta weekend, combined with record numbers of people on the trail for this time of year, have rendered the town *completo*. That word most dreaded by exhausted walkers with no reservation.

The narrow street opens up to a public square. Two lines of motorcycles facing each other, form an aisle in front of the large doors of a church on the right. Their engines idle noisily, all the riders wearing their leather finest. I don't have to wait long to find out what's going on. A bride and groom exit the church and run the gauntlet, to much hullabaloo and confetti from the cyclists. I don't stop to see how this fairy tale ends, as I hear, "MARY JO," yelled above the din.

Although the wrong name, my intuition tells me this is directed at me. I turn to see Rowena, the mom proficient with the F-word, and her daughter Rosa, with the mad hiking skills. Rowena's sitting on a bench, her daughter hovering nearby tapping on her phone, doesn't look up. She smiles nervously and beckons me over.

"Do you have a place to sleep tonight?" she blurts. I see right away she's in survival mode and I give her a pass on polite chit-chat. The worried look on her face stops me from correcting her on my name too, as she clearly has bigger fish to fry.

"Oh no!" I say, "I heard there was a shortage of beds, did you not make a reservation?" I had just talked to her this morning, we stayed in the same village last night. She was waiting for a bus and showed me a scary looking blister covering half the bottom of her foot.

"No, we're calling all the hotels and albergues in the guide, but they're all *completo*."

I wonder briefly if I should offer to let her and her daughter stay in my room tonight, I'm sure they would be happy to camp out on the floor in their sleeping bags. But before I can ask, I turn around and see Charlotte's back disappearing into a large crowd. She's on a mission and waits for no one. I panic, my brain shuts down, suddenly Charlotte has become my lifeline. Now I'm the one in survival mode and cannot make a decision.

"I'm so sorry, I have to go, my friends are leaving," I say as I rudely turn my back on these needy, other friends.

"Good luck," I lamely shout over my shoulder.

Five minutes later we are walking past the Cathedral, a short distance from our hotels. Through the mayhem on the crowded street I hear someone yell, "ANGELA!" – Well, at least this one got my name right.

Aaand it's Kevin from Kona. We hug Hawaii style, like old friends.

"What are you doing here?" I say, pleased but surprised to see him again. I know he has a more ambitious schedule than me, I expected he was long gone.

"My leg." His shoulders slump, he sighs, and his smile dims a little, "It keeps swelling up, and I need to keep resting it – so depressing."

"Do you have a bed tonight?" I ask, knowing this is the question of the hour.

"No, that's the other thing – I've been to all the albergues and a couple of hotels, I've talked to other people, there's nothing left. I considered walking on, but I'm sure by now everything is *completo* even in the next town."

Yes, I'm sure he's right about that, most places are booked by 2pm right now because of the high volume of pilgrims, even without Fiesta time. With a mischievous grin he adds, "I did find a good doorway though, it's pretty well sheltered. If nothing else comes up, I'll go back there after dark and set up my sleeping bag. If someone doesn't beat me to it that is." He sounds like a seasoned, homeless person.

I have a visceral memory from two weeks ago, of that moment when I realized we would have to sleep on a cold concrete floor for the night – part loneliness, part fear, and one psychotic part, adventure. My mind takes a moment to process the only kind thing to do, and in the spirit of Aloha and the Camino...

"I have a hotel room tonight, if there are two beds, you're welcome to share it with me." I blurt rapidly before I can reconsider.

Oh my God! What was I thinking? I immediately second guess myself. But at this moment, everything in my life is so upside down from normal, what's one more thing.

The relief and gratitude that sweep across his face, tell me I have done the right thing. *I have done the right thing, haven't I?* I swallow that thought, he seems kind and gentle. I pray that for once my assessment of character has not been swayed by my alter ego, Pollyanna.

As he gathers his things, I relay the cliff notes of our conversation to Charlotte and Harriett. Charlotte assures me she would have done the same thing. I rather doubt that, but it sounds like she understands my motives, and doesn't think I randomly pick up strange men and invite them back to my room. This is part of my ever-present, *worrying about what other people think,* stuff. It can be exhausting!

I bid farewell to Charlotte and Harriett at their hotel across from the Cathedral. Kevin and I walk a short distance further to the Abba Hotel. To break the ice a little, I tell him:

"Yesterday, when I booked this room, they put me on hold for a minute to the instrumental version of *Dancing Queen.*" I hum a few bars of it – he chuckles.

"I know, I was still laughing when they came back on the phone." We both laugh, then continue to walk on in companionable silence.

The impeccably-dressed gay man behind the front desk of this upscale hotel has a bored, expressionless look on his face. It says, *I've seen it all, and don't really care why you're allowing a strange man to stay in your room.* This, in response to my over-explaining why I now need a room with two beds. I can almost hear his, *mm hmm,* as he kindly shuffles some rooms around. He hands me two key cards and directs us to the furthest freakin' room from the front desk, in an older part of the hotel due for refurbishment. Understand that at this point, my feet are on fire and every step counts.

I slide the card in the door. Kevin follows me into the room, and we both stand just inside the door silently contemplating the arrangement. There's a beautiful green view from the window, but that's not what has our attention. There are indeed two twin beds, but they're pushed together, no nightstand in between like American hotels. I find out later they call this a *matrimonial room,* haha – but I'm not laughing now.

After an awkward silence, Kevin clears his throat, "which bed do you want?" he says. I randomly choose the one closest to the window and sink onto it. My feet are so painful, I don't feel they can support me another moment.

Kevin drops his backpack at the foot of the other bed. He seems unsure what to do next. I'm physically wrung out, in pain and with a confused slew of thoughts running through my head. Normally I would gloss this over with a false sense of bravado and cheerfulness, in an effort to make both of us feel more comfortable. At the very least I'd have a funny quip for the occasion. But I can't seem to summon up the strength. The Camino has systematically stripped me of many of my social graces. Leaving me with only raw honesty.

I remove my boots and it gets better. I'm dismayed to find that the big toe, on the most-sore foot, is white and numb! "Oh crap!"

"Hmm – that doesn't look good," Kevin states the obvious. "I just went through that with my leg – bummer!" By that, I know he means he just had an injury that was obviously going to lay him up for a few days.

"I'm gonna go find some food and leave you to take a bath," he adds. Thankfully he remembers me telling him I have been looking forward to that since yesterday, and he's going to give me some privacy. Looks like I made a good call on *him* anyway.

I luxuriate in the hottest water I can stand, looking at my big toe which is still white and numb, and I reflect. I didn't really want to spend an extra day in Burgos, but clearly, I will need to. I smell a learning here. My strong will appears to be up against my common sense. I like to think I can make anything happen, and that I have control over my life and body. But there is obviously a higher power with its own agenda at work here, and I will need to surrender to it. For the past two weeks, I have been re-asserting that I'm open to whatever experience is in my highest good, as I continually get knocked off my center and have to regroup.

With a sigh, I sink beneath the hot water, looking at my traitorous big toe and ponder my choices. I naively thought I came on this trip free of expectations, wide open to all possibilities, welcoming adventure and the unknown. Ha! How often do I fool myself with those lofty thoughts? When secretly, I'm hoping they will be cushioned by a semblance of familiarity and security.

There are so many ways to *walk your own Camino*, but it never occurred to me that I wouldn't do every step on foot. I now have to make peace with the distinct possibility that I will need to look at alternatives, as my body teaches me to chill out and let go of control. To get to Santiago in time to catch my flight home, I may have to become an occasional *taxi whore*. An endearing term my future friend Rene will call me. Her comic, not so subtle nod to all the judgment, of self and others, that we observed on the Camino.

Kevin returns, and it's my turn to leave to find food while he bathes. I meet Charlotte and Harriett for dinner. It's become obvious to us all that my time with them is winding down. Even though we all plan to stay another day in Burgos, none of us suggests walking on together after that.

That night Kevin and I lay propped up on pillows in bed, inches apart, clean, fed and feeling better, looking for all the world like a *matrimonial* couple. Now that I've determined he's safe, I can smile at this surreal situation.

"I was a Jehovah's Witness for 25 years," says Kevin after a short silence.

I turn to him, "Really?" I'm so surprised I'm at a loss for words.

This sparks a deep conversation about religion and spirituality. The Camino doesn't lend itself to small talk. The shared experience creates an intimate camaraderie. We continue to lay bare our souls to each other, talking late into the night – ok, well, until 11pm anyway – past my bedtime.

Day 17: Rest day in Burgos

It's still dark when I'm awoken by Kevin quietly rustling in his backpack, preparing to leave. An uncharitable wave of jealousy consumes me, as I remember I have to stay put today. When you join the throng of 200,000 pilgrims that walk the Camino each year, you become swept up in a compelling, invisible force that draws you forward regardless of the physical toll on your body. It takes as much conscious mental strength to stop for a day, as it does to struggle through in pain.

"How much for my half of the room last night?" he asks quietly.

"50 euros?" I say, as I search his face to make sure this won't break the bank. It's actually a little more than that, but this was never about the money, and I know he's being more frugal than I am.

"Thanks again for letting me crash here," he adds with a genuine smile as he hands me the cash. "I can't imagine what I would have done otherwise."

I miss the hug that we would surely have had if I were not in bed wearing only a T-shirt. Then, as quickly as he came into my life, he is gone once more. I doubt I will see him again, he has an ambitious schedule. As he closes the door behind him, I want to shout, "Wait for meeeeee," then leap out of bed, magically healed, and walk with him for a day. But instead, I struggle to the bathroom on very sore, unsteady feet, accepting my fate. I'm stuck being a tourist today.

By 8 o'clock I can no longer stay in my room if I want to keep my sanity. Determined to try and make the most of this rest day, I enter the square under blue skies, with a slight nip in the air, in search of a leisurely breakfast. This should not be a problem, as nobody rushes you here. Paying the check is a toe-tapping lesson in patience, often requiring you to chase down the waiter, waving money furiously at him. I find I am eating much slower these days also, enjoying my food in a new way, one bite at a time.

"That looks like a decadent breakfast," I say to the English-speaking older couple sitting at the table next to me. They're sharing a plate of hot churros and chocolate sauce. While they look like tourists and not pilgrims, the shared experience of being out of our element in a foreign country breeds casual intimacy.

"It bloody well is," the gentleman replies, patting his tummy with a mischievous grin. "I couldn't do this every day and keep my trim waistline."

At another table is Mr. Dark and Broody, the flirty German man, and his son that I met the other night at dinner. Neither of us is in the mood for flirting this morning however, so a cursory nod and smile are sufficient. I will not see them again.

Standing in front of the Cathedral, dwarfed by its immensity, I notice a large red cross next to a small, ordinary-looking door which is invitingly open. A steady stream of people enter through it. As I am not a Catholic, I was unaware until after I committed to this trip, that the Pope had declared this a special Holy Year for Mercy. I now realize that this unassuming door is a Holy Door. Only a few select cathedrals around the world have these and two are on the Camino, the other is in Santiago. They are only open during Holy Years, and if you enter through one, you are considered absolved of your sins, or something like that. After December they will be closed again until 2021, and after that another 25 years. So, without further ado, I walk through it.

Well, that was a little anti-climactic. I'm not exactly sure what I was expecting, but I'm not interested in sitting through a service. Since that seems to be my only option, I walk back through it and around to the opposite side of the Cathedral to the main entrance. Here I can sign up to take a three-euro tour. As I'm fiddling with my audio set, I become confused, go the wrong way and end up doing the tour backwards. I of course have to abandon the audio set. One of my daughters says "shocking!" when we laugh about this later.

The artwork, gold, sacred geometry, the crypts where famous kings and holy men are buried (including El Cid), the stained glass and gothic architecture, are absolutely mind-blowing. I sit on a bench looking up at the middle spire which is several stories high and do the math. How did they do all this with no technology or machines? It took several centuries to build, starting in 1021. This means a long series of workmen, spanning many lifetimes, had to hold the same vision, without blueprints as we know them – my mind is officially boggled.

As I turn my headset in and prepare to leave, a deafening, magnificent sound stops me in my tracks. A chorus of church bells in a range of notes, are ringing a glorious, complicated song. It continues for 15 minutes. I walk around and sit on a bench in the square, across from the Holy Door, to bear witness. Tears come to my eyes at the sheer beauty of this unexpected gift, I feel well and truly blessed.

Something catches my eye. I turn to see Mary Jo and David with their backpacks, moving swiftly on the opposite side of the square. They walk purposefully, obviously leaving town. I begin to raise my hand and consider calling out to them – but hesitate. Mary Jo, dwarfed by her pack is in the lead, head down, not paying attention to David who is ten feet behind her, also head down. Neither of them seems to be aware of the miracle of the bells. I have this fleeting flash of them as an old married couple, that no longer have anything to say to each other. The kind you see out at dinner, eating an entire meal together, whose only conversation is, "please pass the salt."

I know I'm making this up, but my angels are using this to remind me of relationships I've stayed in a little too long, and they've ceased to be relevant. The writing was on the wall, but I wasn't quite ready to let go of the comfort and false sense of security, even a dying relationship provides. Hah! Kind of like the relationship I had with these two! I chuckle to myself – now I know what to do.

The bells are still ringing, I doubt they'd hear me anyway.

Day 18: Hornillos del Camino

The scent of coffee wafts under my nose as I walk through the lobby of my posh hotel at 7 a.m. this morning. I almost make it to the revolving door. I have become used to waiting for several kilometers for my first cup, and I'm anxious to get going, but the smell does me in. After a moment's hesitation, I turn on my heel and head towards the sound of gentle clinking of silverware on china.

My dusty boots and large backpack feel out of place in the white-linen-tablecloth dining room. The aroma of bacon and sausage mingle with the tantalizing smell of coffee. A combination that can still make my mouth water, even after decades of not eating pork. A tall, slender woman in a business suit and stiletto heels barely gives me a glance as she chooses her breakfast from the extensive buffet.

I just want a café con leche, this is all too much. I turn around ready to abort mission, but the hostess is standing right behind me. She immediately grasps my dilemma. With a gentle smile, a head nod and a game-show-hostess hand gesture, she gracefully points to the first, table-for-two, barely inside the door. Ok, I can do this, sitting here I will not have to walk past several groups of smartly dressed people.

I mumble apologetically, "Gracias, umm, un banana y café con leche por favor?" I've forgotten what the Spanish word is for banana, so I say it in English with an accent. Ha! Like when someone who doesn't speak your language yells at you as if you're deaf. Because of course, saying it louder, or with an accent will suddenly make it recognizable. But she seems to understand. We have conducted this entire transaction without her uttering a single word.

The banana has seen better days, but I eat it with gratitude. Finding fresh fruit is always a bonus. As I finish, I catch the hostess's eye and offer my money, but she shakes her head and waves her hand in front of her face and with a smile wishes me, "Buen Camino." My eyes water as I swallow hard not to burst into tears. Random acts of kindness are legendary along the Camino. I put my hand over my heart and return her smile with a, "Gracias." This feels like a good omen, and indeed it is. The memory of this will get me through the next couple of hours.

It appears that pain will be my constant companion today. I optimistically set out thinking my feet are better, but it quickly becomes apparent that I am wrong. I slow my pace, take smaller steps, and that seems to help. I begin repeating, "My feet heal and feel better with every step." But miracles seldom repeat themselves, and this is one of those times.

Fortunately, the walk out of the City is quite beautiful and short, compared to the torturous trek in. I focus my attention on the pretty, brick path bordered by trees in full bloom. I'm passed by one, then three more solo pilgrims, we acknowledge each other with a quiet, "Buen Camino." I walk through a neighborhood of two-story condos, after which the Camino goes from concrete sidewalks to a gravel, country path between fields. This is all made more beautiful by the silence and the morning mist bordering on fog, still hovering over the tall grass. My mind is suddenly crystal clear, I feel fully alive. All my senses are drinking in this moment, there is nothing but Now. I have a full body experience of total peace.

I'm so glad I stopped for that first coffee, as the next town is an easy, but long, 12 kilometers. With a sigh of relief, I cross the street and enter a café with pilgrims gathered outside. The three older men sitting at the bar drinking their café con leche, turn to glance at me, then go back to reading their papers. The young, dark-haired man serving behind the bar is surly and takes my order with no comment. I'm not sure how I'm responsible for his lousy morning, but it's not going to spoil mine, I give him the most genuine smile I can muster.

Outside, I finally sit with a big sigh. I remove my boots and put my most painful foot on a chair. I take a bite of a delicious *bocadillo* (sandwich) with tuna, egg, tomato, and lettuce on some crazy-good bread. My taste buds are even more alive this morning, as the different flavors and textures move around my mouth. I rise to leave and to my immense relief, my feet feel a lot better. The brief rest was good, note to self (which I will immediately forget): Rest more frequently.

There is something special about this next tiny village. I don't have time to pause and put my finger on it. The light is hitting it in a certain way, there's a peacefulness begging me to rest a while, but I don't. Whatever it is, I make a mental note that if I'm ever crazy enough to do this again, I will stop for a night here in Rabe.

There's no billboard or marching band to let you know that you've officially entered the Meseta. Just a long, steady climb to the flat plains of Northern Spain. I will spend a week traversing them. Many say that the Camino is split into three sections. The first that I have just completed is the physical. During this part, your body is hopefully getting used to all the walking. Also, the terrain is quite challenging with a lot of elevation changes. The Meseta is the mental. This is very flat composed of mostly wheat fields as far as the eye can see. So, it is an excellent opportunity to *go inside*, without too many distractions. The third portion, Galicia, (pronounced as if you have a lisp, Galithia) is the spiritual. A nature walk that is a testament to Gods creative abilities, and an opportunity to gather everything you have learned from the trip so far.

I have been walking for several hours, and according to the mileage in the books and the GPS on my phone, I should be at my destination. But all I see are green fields of lush grain, going on forever, bisected by a now familiar wide, white gravel path. "What the...?" I mutter to myself. I'm still in the middle of nowhere. I wonder how I will continue on these pitiful feet, but I have no alternative. None of the distances ever seem to add up, it changes from one source to another. The Camino defies modern technology, it begs you to put all your books and apps away and just walk.

Finally, I crest a hill a kilometer further on, and there below me is Hornillos, the sweet little village that will host me tonight. I could drop to my knees in gratitude. After a short distance I come upon the private albergue I booked for this evening. It already has a sign on the door, *Completo*, for those with no reservation.

I take a short rest and shower, then set off to explore on this gorgeous afternoon. I have walked no more than 30 ft from the front door when I make a U-turn. My feet are screaming at me to stay put. I decide instead to take up residence in one of the lounge chairs in the grassy backyard of the albergue. I close my eyes, and with a deep sigh, surrender to the warmth of the afternoon sun. Soon, I hear several posh, English-accented voices heading my way.

"Well, what am I supposed to do with these dirty socks?"

"Bloody cheeky if you ask me, does she think you're going to hand-deliver them to her?" Somebody else replies indignantly.

I peek to see three people taking up the remaining chairs around me. A man and woman in their 60's, and another woman at least ten years their junior. The younger one, whose name I learn is Sophie, is attractive, slender and has the body of an athlete. She is the bearer of the dirty socks, that I later learn belong to another travel-mate not staying at our albergue. The older woman Victoria is making it perfectly clear, that she will not be taking responsibility for them.

Victoria is the archetype of a past-middle-aged Englishwoman. Typical in a nondescript kind of way. Slightly overweight, with short, curly hair and a stern, unsmiling expression. I would guess she is the matriarch of the group.

She is traveling with her husband Colin, who on first impression appears to be the quiet, submissive type. Victoria is pouring over the guidebook, trying to decide where to reserve for the next couple of nights.

"Well, we have lots of options in Castrojeriz, here's a rather nice-looking place for only ten euros," she mutters.

Meanwhile, Sophie is content to let Victoria make the decisions about lodging, while she continues to fret about the socks.

"...and they're wet and dirty, does she expect me to wash them too?"

Deciding to stretch my social muscles, I open my eyes and join in the conversation, as this is what people do on the Camino.

"So, where are you guys from?"

Well, it's what most people do. For a moment the silence is deafening. But Sophie is tired of obsessing about someone else's footwear for now.

"Cambridge, England," she replies, with a slight upturn of her chin and a brittle smile.

Victoria however, glances away from her book momentarily. She looks down her nose and over her glasses at me, the intruder, and without saying a word expresses *WHO are YOU? And why are you talking to us?*

I chuckle to myself, she's not going to intimidate me. I was born in England, I know how this works. A presumptuous American is beneath her attention, my use of the word *guys* probably outed me. But a fellow Brit – well, she might be given the time of day.

So, I quickly chime in, "I was born in Cheltenham, we moved to the States when I was 13."

"Oh, Cheltenham's lovely," says Sophie "where do you live now?"

"In Hawaii."

You have to hand it to the British. For centuries they've felt superior to the rest of the world and have perfected their *I'm so not impressed* look, and that's what I get now. Colin is studiously reading a novel and ignoring this whole discussion.

While me being from Hawaii is a conversation starter with most pilgrims, the subject never comes up again. I intuitively know that being born in England is my ace up the sleeve with this group. I've already played that card though, so I'll just leave it on the table and see what happens. Sophie goes back to fiddling with hanging the offending socks over a rail on a drying rack.

With a little sniff, a "hmmph" and a flicker of her eyebrows, Victoria returns to her guidebook. She still has not said a word to me, but I can see that she has softened somewhat. For someone of Victoria's station (from her accent I can tell she's upper class), in the grand hierarchy of things, English blood still counts for something. In this time for me of vulnerability, loneliness, and uncertainty, their very Englishness feels familiar and comforting. With just a few minor adjustments, Victoria could be my mother.

During the evening meal, a wonderful homemade paella, we all sit together telling stories and laughing. Victoria starts to warm up to me and seems to grudgingly decide I'm ok. I don't for a minute mistake this as an invitation to join their group, just that I'll do as a dinner companion.

At supper I meet June, the fourth member of their party. She is also in her 60's, blonde, cute, funny and lighthearted. The fifth person, Kate-of-the-dirty-socks, is staying somewhere else.

"I just lost my husband of 40 years and we used to do everything together. I decided it's time to start doing things on my own," June tells me.

"Go big or go home," I say. After a few seconds, a flicker of comprehension passes over her face and she smiles, as she deciphers the unfamiliar saying. Her blisters have been so bad, that while they heal, she's had to take a bus or taxi for five days to keep up with everyone in her tribe. This *weakness* has put her at the bottom of the pecking order, in what I am starting to learn, is a highly stratified group.

This charming, funny, opinionated and competitive, assortment of characters, will continue to intersect the remainder of my journey to Santiago.

Pictures

The Pyrenees

Fairy Bridges

Crazy stairs

Steep Streets

The Wine Fountain

The long and winding road

Kevin leaving Cirueña

The Meseta

Annie

Leaving Castrojeriz

Kevin and I in Fromista

To the Cruz de Ferro

Cuckoos and frogs

A roman bridge

Country lanes

Day 19: Castrojeriz

This morning I decide to send a bag ahead to my next destination with 10 lbs of my gear in it, leaving me to carry 8 lbs. To give my feet a break, it was; ship a bag or ship me. I'm having a little pity party as I start the day once again in pain, causing me to walk at a snail's pace.

I am passed early on by the English cohort, with pitiful glances at my weakness, and a mumbled, "Buen Camino." *They don't think I'm going to make it,* I think to myself. In their world it is survival of the fittest, the weak are left behind for the predators and are to be pitied. They're followed shortly by two German women I met last night in the surrounding bunks. After giving me looks of concern, they disappear ahead of me on the beautiful country path between fields, with a cheerful, *you can do this,* "Buen Camino." I choose to ride their wave of optimism.

The mystical, foggy morning makes other pilgrims impossible to see and contributes to my feeling of aloneness. Oh yeah, *the Meseta* I think to myself – *Ahhhh! Time to go inside.*

A thought floats into my mind. *What is this pain about?*

I pray for God to please take this burden whatever it is. I burst into tears, sobbing with abandon, there is no one nearby to hear. Within minutes my feet feel tremendously better. Half an hour later the pain is mostly gone. Whether it's due to the emotional, endorphin release or the ibuprofen kicking in, or a combination of both, I don't know or care.

As my funk lifts so does the fog, to reveal another gorgeous, sunny, walking day. Gentle hills and long, wide-white paths between wheat fields stretch to the horizon as I pick up my speed, almost euphoric at being able to walk normally again.

The first rest stop after ten kilometers, Hontanas, was built in a dip in the landscape and seems to appear out of nowhere. Just before entering the village is a tiny building with an open wrought iron gate for entry. I duck to clear the doorway, and in the dim interior, I see a statue of a saint. I cannot understand the Spanish write-up next to it but decide this feels like the right spot for my Cooper prayer today.

I line up behind Victoria at the first bar, for my café con leche and breakfast. She turns, visibly surprised and asks accusingly, "How did you catch up with us?"

"I walked," I say with a grin. Then, deciding not to push it, I add, "My feet started feeling better, so I was able to walk faster."

I join them at their table outside, Sophie has taken off her shoes and is doing some stretches.

"You should do this every morning before you get out of bed," she says as she stretches her feet up and down, then side to side then rolls her ankles first in one direction, then the other.

"Hmm, yeah, that makes sense."

"Here, take off your shoes, try it now," she orders

I comply. "Where did you learn to do this?"

"Oh, I run marathons, these exercises have saved my feet and ankles."

I will do this drill every day, before getting out of bed, for the remainder of my Camino. As we sit there, June hobbles by on her blistered appendages, with that familiar rolling gait of someone walking in pain. She continues on, smiling bravely at us as she passes our table.

"You should stop and rest," Sophie calls out to her. But with a shake of her head and a cheeriness that I'm not buying, June says, "no, I'm ok, I'll meet you at the next town."

I'm familiar with this tactic, I discovered it over the past week. When you have to walk slowly due to pain, you can't take frequent or long, rest breaks, even though you need them now more than ever. There is a strong, primal urge to keep up with the pack. If you fall behind, you are awash in a lonely sea of strangers, having to start your social affiliations all over again. I found this becomes more difficult as the Camino progresses, because many people have already formed *Camino families* and are not as open to admitting others. In June's case, she came with her *family* and needs to keep up or take a taxi. In her competitive group she is up against a subtle, *failure* vibe, for physical weakness. She is definitely low in the pecking order of their hierarchy.

I stop to say a quick hello to Charlotte and Harriett, my Aussie friends, stationed at a nearby table. Then wave at the two German women who passed me earlier on, sitting at another table. They are all surprised to see me. What an illusion life can be. I started off three hours ago feeling deep aloneness, and suddenly I am surrounded by all these international friends, at this beehive of pilgrim activity.

After eating, I leave before everyone else, as I need to get to Castrojeriz before the farmacia closes for siesta to get some Voltaren, a magical ibuprofen infused cream. Some was given to me last night by the hospitalero at the albergue, and definitely contributed to my miraculous recovery this morning. In my euphoric, pain-free state, the path seems to sparkle, wildflowers abound. I take lots of pictures for my grandson Cooper, who loves flowers more than pizza and ice cream. I send him one or two each day to briefly lift the wet blanket of his chemotherapy-laden day. I'm totally alone on this beautiful trail as it winds around the mountain. To think I was afraid of this, being by myself in the middle of nowhere. It turns out to be the ultimate freedom.

I catch up with June stoically limping along. She is cheerful in spite of the pain I know she is in. I slow my pace and we chat for a while, but I feel an urgency to keep going. If I don't get to the farmacia before siesta, when everything closes down until 7 p.m., it may be several days before I get to another town with one. I don't relish the thought of being in that much pain again. Between the industrial strength pills and now the cream, I seem to have found the magic combination to keep me walking. I can't think about how much ibuprofen my liver and kidneys are having to process. I'll need to do a cleanse when I get home.

I bid June, "Buen Camino." It secretly feels a little good that I'm the one passing another pilgrim today. Shortly, the trail takes me to a tree-lined, rural road. Now pilgrims seem to appear out of nowhere, one at a time. I go around a corner and see the ruins of an old monastery up ahead. At first, I get my hopes up thinking this is my *Hallelujah moment*. I try to stifle my disappointment as I walk under a stone structure that arches across the road and wonder at the beautiful old architecture. My nose wrinkles and brow furrows as I smell incense, and then hear new age music. A wave of familiarity washes over me. I chuckle out loud when I see the sign in a window for a massage therapist. Haha, how predictable! I guess we are the same the world over.

With a sigh, I now have my *Hallelujah moment*. But it's still three kilometers in the distance. The clock is ticking, and my ibuprofen is wearing off.

Castrojeriz is a larger town than I thought, and of course, my albergue and the farmacia are more than a kilometer away on the other side of town. I walk up the inevitable hilly streets following other pilgrims, in search of my brand new albergue that no one seems to have heard of. I stumble upon it with 15 minutes to spare. I drop my bags on a single bed by an open window, in a room with only four beds, practically a private room. Then I race to the farmacia.

It's time to get caught up with wash, and this albergue has the luxury of a washing machine, and a small, sunny courtyard to hang the clothes in. Friendships on the Camino form hard and fast. One of the German women I bonded with over sore feet last night, is also staying in my room. She asks if I'd like to join her and three other German friends for dinner. *Jawohl!* There, we find a common spiritual interest that goes beyond injuries, and probe a deeper connection, as best we can in the group. The others try gallantly to politely include me in conversation, but only two of them speak English. They invite another German man, dining alone, to join us. At the end of the meal, as an afterthought, he asks everyone's name.

As I return to my room, *ting*, there's a message from Kevin. Since our stay together in Burgos, we have been in touch daily through FB messenger. We continue this the remainder of the journey. I think we all want someone to witness our life, especially a trip as momentous as this.

"I'm in Fromista, I'm taking the day off tomorrow, my leg's bothering me again. There's a nice little hotel next door, do you want to come here and share a room together again?"

I reply, "I'm in! See you tomorrow."

Fromista was my planned destination anyway. Staying at the albergues is inexpensive and offers authenticity, companionship and the opportunity to meet new people and a certain amount of adventure, while hotels provide comfort. I'm finding a hot bath once a week to be very therapeutic for body and mind. Since the purists often turn up their nose at comfort on the Camino, and Kevin appeared to be one of them, it would now seem that I've drawn him over to the dark side.

Day 20: Fromista

The weather Gods love me! The quiet walk out of town, through a mist hovering above the fields and the fresh smell of dew on the grass, leaves me feeling suspended in time. The sparse line of pilgrims adheres to an unspoken code of not talking in the golden, morning hours – just a little head nod and a soft, "Buen Camino" from those passing you.

We soon face a fairly steep, long climb on a trail going around a small mountain. Although I walk alone, I use the pronoun *we* sometimes, as there is a sense of being part of a larger community. In single file, or in small groups we follow each other.

From the top of the hill, Castrojeriz looks like an elaborate Lego town. It is stunning bathed in the early morning light. However, as I crest the ridge and look forward, it is for me the quintessential Camino picture, both on my phone and in my head. The wide concrete path descends steeply to the open arms of green, wheat fields, and a gravel trail snaking its way to the horizon. Pilgrims up ahead become smaller and smaller, until the colorful specks of their backpacks are no longer visible, only the endless trail.

I see why they put concrete on the first part of this steep, downward path. Due to its angle, the only way to traverse it safely, and save my knees, is to zig-zag. An idea I poach from the pilgrims walking in front of me. My feet are forcing me to go slowly, but I am perversely excited about this long walk today, seemingly to nowhere.

For me, this is the Meseta at its finest. Eight kilometers with no sign of civilization is a really long way. The path is as beautiful in its starkness, as it is boring in its sameness. An enterprising local has set up a little oasis, where after two hours of walking, I stop to sit. I put a euro in his donation box in exchange for a styrofoam cup of coffee, thick and black. Adding milk and sugar doesn't help, it's disgusting. Even though I'm desperately in need of some caffeine, I just can't do it. I dump it. I've already learned that you don't always know where the next rest stop will be, but I'll have to take my chances.

It turns out I don't have to wait too long this time. After walking for two more kilometers, the small hamlet of Itero de la Vega appears around a bend in the trail. The first building is a bar with a large outdoor seating area filled with pilgrims, laughing and chatting. The sun illuminates their colorful clothing. I experience sheer joy at the simple prospect of a real breakfast and an excuse to sit for 15 minutes. A Danish couple I met a few days ago wave me over to join them.

The path out is ten and a half kilometers of the same, flat, monotonous plains of wheat. Thankfully my angels send along a distraction. Fred from Georgia is a curmudgeonly old man. He is bent over his walking sticks, with a bow-legged, rolling gait, and entertains me with his version of how the Camino should be walked.

"All these damn kids with sayll (cell) phones," he grumbles in a delightful, southern accent. "They call ahayd and book up all the bayds, it takes me longer to get where ah'm goin' 'cause of my age, then there are no rooms available – shouldn't be allowed."

Unconsciously, I put my hand on the hip bag in front of me, that holds my *sayll* phone. I can't quite keep the slight smile off my face as I ask, "have you thought of asking at the albergue you're staying at to book ahead for you?"
"That's not hayow you're supposed to walk the Camino," he drawls.

Ok, now we're getting to it – "What's the right way?" I ask innocently. I can already see where this is going – but we have ten and a half more kilometers, this will keep my mind off my feet. Also, I'm amused to hear him say out loud, what so many think quietly to themselves.

He takes the bait, "Everyone should wahlk and carry their own pack. When they're done for the day, they stop and fahnd a room. It's not fair when people like me are turned away, I cayn't help it that I'm old and cayn't walk fast, they should save rooms for people like

119

me, who are slower but doing it the *raaght* way. The old-time pilgrims didn't have sayll phones, and there were no taxis," he adds.

I resist saying, *Yeah, and they walked with a piece of leather strapped to their feet if they were lucky, and many of them died from injuries or were murdered by bandits. There are a lot of things different in today's world.* But I suspect he wouldn't have cared about this perspective, as it didn't support his narrow view.

Changing the subject, he says, "So, you're walking by yoursayllf?"

"I started with 4 other women, but for various reasons, we all went off on our own."

"Oh, are you with Mary Jo's group?" He asks

"Yeah, how did you know about that?" My eyes are wide with surprise.

"She's on the email forum ah'm on, I remember her asking for female walkers."

Ha! I stifle a laugh at that coincidence and wonder at what a small world it is, even here on the other side of the globe. "Yes, she's a couple of days ahead of me now." No need to get into that drama, even though I suspect he would have lapped it up and it would have made his day.

"Wayll, you tell her I sayd hello when you run into her." For the time being, he's forgotten his Camino rules and complaints.

We walk in silence for a while, listening only to the crunch of gravel under our feet and the clicking of our poles. Like tuning forks, we unconsciously match our rhythm. I think back to all the subtle expectations I had for this trip too, some not unlike Bob's. I fancied myself as a *true* pilgrim, whatever the hell that is, also. I guess he's here to hold that mirror up for me and bring me to my knees once again. Right now, I'm just getting through each day the best I can, and watching my expectations roll out of reach in every direction, like a dropped basket of apples. Attempting to catch them as they tumble away is a comical exercise in futility. In my experience, the Camino in all its glory, levels the playing field. Who cares if you're a lawyer or a janitor – how many miles did you hike today? What do you do for blisters? Have you had to poop in the woods yet?

As we approach Boadilla del Camino, my feet are so sore I'm not sure I can walk the remaining half a kilometer. Even old Fred has moved out ahead of me. The urge to lay down and curl up and cry in the wheat at the side of the trail is almost overwhelming. I remind myself of the many species of poisonous snakes in Spain that live in the tall grass, to keep me moving forward. One, two, three – count the steps, focus on using the poles correctly and breathe. *You can do this.*

What a run down, unappealing little place this is. Although I can walk no further, nothing in me wants to stay here. The prospect of a hotel room, a hot bath and an evening of friendship with Kevin at the next village, motivates me to let go of one of my most cherished rules for this trip. I break down and call a taxi. I watch Fred's back shuffling down the street, glad that he, of all people, will not be around to witness my shame.

Perhaps Fred was sent by my angels, to help me see how ridiculous all of our self-imposed regulations are. We make them up using information we have at the time. For me this was comfortably sitting in Hawaii with a cup of tea, on my lanai reading books, far removed from the actual experience. Then we act on them as if they're real and forget to re-evaluate based on new circumstances.

I ask to use the restroom at the dilapidated Municipal albergue, which makes me wish I'd found a spot somewhere behind a tree, snakes or no snakes. When I ask for the phone number of a taxi for the remaining five kilometers, the woman behind the bar drips disdain as she points to a small business card tacked to the wall – pilgrims are not the only ones judging each other here.

What a welcome sight Kevin's smiling face is. He hugs me warmly and I wonder, not for the first time, if there might be something more between us. I don't think circumstances on the Camino (tired, smelly bodies) exactly lend themselves to romantic relationships. It could be it's just been so long, it doesn't occur to me anymore. Rene in one of her Buddha-like moments, looked out at two young lovers one day and remarked, "how do they have the energy to fuck after walking all day?"

While not grand like the Abba, this small hotel is as quaint and well-kept as an old lady dressed for church on Sunday. Kevin nods to the man behind the front desk who has noted my arrival, and I follow him up a flight of stairs. I have déjà vu as we stand in the doorway looking at the two beds, matrimonial style. Kevin and I look at each other, we both smirk, no words necessary. He's saved the bed furthest from the door for me – he remembered – then tells me to enjoy the tub, he'll meet me outside. The door in my heart, creaks open another notch.

I know I should get horizontal for a while following my long, hot bath. But the brilliant sunshine coming in through the screen-less, wide-open window beckons me. I cannot stay inside on such a gorgeous day.

I take a sip of my sangria as I sit under an umbrella and put my bare feet up on the opposite chair. I need to write my blog, or *my people* become disappointed and wonder if I fell in a ravine. On day six, the WiFi signal was so weak I couldn't upload anything except text, so I just sent that. The next day there was an uproar, at least four people wanted to know where the pictures were. So, I had extra homework that night. I've never written a blog before and hadn't realized what a big responsibility it would turn out to be. But without it, how could I possibly convey to my family and friends who really care, the magic and vastness of my experiences here?

After any trip people will ask, "so how was your trip?" Seldom do either of you have the time it takes to properly answer that question. A blog seemed like the perfect tool for people to follow me in real time.

It turns out my commitment to my blog was also a commitment to myself. Without it, I could not have organized the memories of my trip in a coherent way so as to write this book. Kevin joins me just as I'm finishing, and I take a selfie of the two of us to add to my blog for the day.

"Well that was pretty boring today. I get so antsy just sitting around. I came here to walk." Kevin voices the frustration we all feel when a day off is mandated by our body, not by choice.

"Do you wanna walk together tomorrow?" he continues. "You go slower than me, so it will force me to take my time and let my leg heal some more."

"Yeah," I say, "be nice to have company."

I enjoy Kevin's easygoing personality. He's easy to be around, easy on the eyes, everything about our friendship is easy. In a flash of insight, I realize the other thing that is so familiar about him. He reminds me a lot of Bill, even looks a bit like him. Bill is one of my ex's that lives in Texas. One that I let get away. He was the first of my trilogy of boyfriends to go off and marry the next woman he met. I sometimes still wonder *what if?* about him. Although, following shortly on the heels of that breakup, was my transformative decision to move to Hawaii, and a life I couldn't have imagined in my wildest dreams.

After dinner, we retire to our matrimonial beds. Propped up on pillows, we look at each other and smile.

"So how was your day dear?" I ask tongue in cheek. We both laugh.

With so few other distractions, one gift of life on the Camino is connecting with people face to face, without the filters of social media. The relationships feel more authentic at some level.

Day 21: Carrion de los Condes

It's my birthday! Kevin and I get an early start and hobble out into blue skies, sunshine and a gentle breeze. Well, *I* hobble, you couldn't tell from his long, effortless stride that anything's bothering him. The path is wide and flat, promising an easy walk today. Thank goodness, because my feet are not doing the happy dance yet.

After several kilometers, a little Hansel and Gretel bar appears in a clearing in some woods. I don't see any houses, but if we are on the outskirts of a tiny village, it would be Poblacion de Campos, the timing is right. We are the first pilgrims this morning to find this short, squat little shack, surrounded by tall trees. A smattering of dead leaves on some of the tables and chairs, give it the faintest air of abandonment. But it has everything we need; an ordering window for food, and a nice bathroom in the back. So, it's an easy decision to stay for breakfast. We unclip our backpacks and set them down with our poles.

I stop and take a moment to breathe in the musty, woodsy air and this luxury of time. I'm getting used to these days where there is no rush to be anywhere or do anything. I'm hoping I can take this back with me. I hear a distant cuckoo, and a memory from long ago takes me back to the bluebell-carpeted woods of my childhood in England. Then, as if on cue, part of the English cohort shows up.

"Victoria, Colin, Sophie this is – ah – Kevin." The introduction feels stilted like I've been caught at something, I feel myself flush a little. What's up with that?

Victoria tilts her chin down, looks from him to me, then raises her eyebrows slightly at me. Sophie looks genuinely puzzled for a moment, her eyes flitting from me to Kevin and back, like, *what's he doing with her?* I'm a little embarrassed at their assumptions and want to explain, but I can see that going downhill faster than an Olympic luge.

"Why don't you join us," I say politely, wanting to move on from this awkwardness.

"Ooo lovely, thanks," says Sophie, and plonks herself down on the other side of Kevin. She turns towards him and gives him a big smile. Oh, I see – it's open season.

While I leave for a few minutes to order my coffee and go to the bathroom, Sophie discovers Kevin's a marathon runner also, piquing her interest even more. When I return, I catch the tail end of the conversation.

"...my last marathon I couldn't finish because I ended up with a bad injury to my leg," says Kevin, "and now it's coming back to haunt me, I've had to take a few days off." He sighs quietly, his shoulders slump a little.

"Well, the last one I ran I placed fifth," Sophie exclaims proudly with a sniff and a little upward tilt of her chin. Her competitive spirit is oblivious to Kevin's disappointment.

Later, as we leave them to finish their breakfast, I tease Kevin, "I think Sophie was a little sweet on you."

"Really?" he says with a puzzled frown. His eyes look down as he tries to recall the conversation. "I don't think so," he finally concludes with a little shake of his head.

"Men are so clueless!" I say, as I laugh and slap his arm.

We walk on quietly for a while, Kevin in front of me on the narrow trail, serenaded by the distinct sound of cuckoos sounding exactly like their clock namesake. The swiftly flowing stream following the path gurgles and rushes, sounding like one of those dreadful Nature CD's you see in the clearance bin at Walmart. As if anything laid down on a little piece of plastic and played in a room with four walls and closed windows, could ever replicate the 3D technicolor version that permeates all your senses. Small arterioles veer off from the aorta of the mainstream, necessitating crossing of several little bridges. We stop on one to listen to frogs bellowing back and forth to each other, impossibly loud.

"Maybe it's their mating call," I say, "It's spring after all."

We come across the English group again at the next bar. They'd passed us a while back. As we stand to wait for yet another café con leche, we are serenaded by a tinny, but haunting version, of Amazing Grace coming from the radio. The voices sound familiar. Then I realize it is Il Divo, an international male-quartet. I love their music. Yes, this birthday does indeed feel grace-filled.

More road walking brings us to Villalcázar, the last village before our destination today. A majestic church looms before us on one side of the square. The pale-yellowish, sandstone exterior seems to glow as it blends with the white brick road and sidewalk. A Flower of Life sculpture on the front of the church catches my eye. It is five feet across and delineates a stained-glass window. The Flower of Life is considered a sacred symbol and is found on many ancient structures around the world, including several in Egypt. Above the arched doorway to the church are more than a dozen figurines etched into the stone, all centered around Mary and Child.

"I have to go in there," I say to Kevin, not a request.

"Okaaay" he says slowly, although I can tell by his response that visiting churches isn't necessarily something he's here to do.

A man behind a small podium requests one euro to enter. Kevin doesn't have one, or doesn't want to spend it on a church, so I pony up two euros for both of us, after all it's my deal.

Standing in front of a statue of Mary, depicted in gold and white, I gaze into her stone eyes and ask for continued protection for Cooper. Kevin walks around pretending to be interested in the stained-glass windows and statues, hidden in various nooks and crannies. He tires of it quickly and goes outside to wait for me. It dawns on me that I have no idea what his thoughts are on mainstream, Christian religions following his long affiliation with the Jehovah's Witnesses.

The last six and a half kilometers are hot, flat and monotonous. A three-foot-tall stone marker, with a blue and yellow scallop shell adorning it, is placed in the middle of the path about every 60 feet. As if the blazing sun will cause us to forget where we are going. The busy road we walk beside the rest of the way sucks our remaining energy.

We have been talking all morning, and I think both of us are craving some quiet time. As the pain in my feet escalates, the way it always seems to towards the end of the day, my pace slows, and I can feel that Kevin is anxious and irritable at the delay. But he's a gentleman and given a choice, he insists he wants to keep walking with me.

As I have not slept in a convent yet, we hope to find space in one of the two in this town. My angels are watching over me. We take the last two beds in the first one we arrive at. To be honest, it doesn't feel like anything special. We never see a nun, just a very cramped room with bunks taking up every available inch of real estate. I have to turn off my claustrophobia filter. I find out later that of course, the other convent was the preferable one, with nuns singing and probably washing your damn feet too.

After showering we set off to explore the town a little and are surprised to find one bar open during Siesta. I have not had a Coca-Cola in several decades but have heard many pilgrims rave about how good they taste on the Camino. For some reason, it sounds really good after the long hot walk today. So, in keeping with busting as many of my rules as possible, I order one and take it to an outside table. I start off being pleased that it comes in an old-fashioned glass bottle. But that is short-lived as it doesn't taste as good as I remember. And a few sips in, I get a picture in my head of a mechanic using it to degrease an engine. Followed with the vision of another burly guy, with hard, calloused hands, pouring it on a trailer hitch to get the rust off. Then I see my delicate, soft, pink intestines – ok, you get the picture. Once you know, you can't not know anymore – I dump the rest. I guess some rules aren't meant to be broken.

Kevin eschews the culinary delight of a Coca-Cola and goes off in search of a nice restaurant to take me out tonight. He's insisting on buying me a birthday dinner. He returns within half an hour, like a cat with cream on his mouth, he's made us a reservation.

We return to the restaurant later that evening and are shown to our table by a waitress wearing a white-shirt and black-skirt. There are indeed expensive ala carte dishes available, but on the back page is the Pilgrim menu. The price disparity is so evident, we both order off the back page. For me, this is more about the ambiance and company, not about setting Kevin back four days on his Camino budget. The scallops did look good though. With the free bottle of wine that comes with all pilgrim meals, we toast to my birthday and our joint gratitude for this amazing experience we're sharing.

We are surrounded by several groups of people we know, including the English posse. Sophie gives Kevin a long look as they are leaving, and me a brittle smile. I have a momentary feeling of belonging to this *other* club. The one where couples are talking to each other over their entree. Instead of the lone diner, reading the menu like it's the last great novel, then turning to her phone as if her life depends on seeing what her friends are eating for dinner tonight on Facebook. In this *other* club you have someone who is pleased and proud to walk into a restaurant with you on their arm. Well – in our case, I was more *clinging* to his arm because my feet were so painful and unsteady. It was necessary so that I didn't trip and fall face forward into the waitress walking by with four dinners balanced from shoulder to hand. In this *other* club at the end of an evening, you have someone to say, "do you want to watch a movie before bed dear?" Instead of, "were you good puppies while mommy was gone?" I haven't belonged to the *other* club in a very long time. In large part because I know there are versions of it that aren't quite as rosy. But this taste of it is surprisingly intoxicating. I could get used to this – I admit that my smile back to Sophie was just a tad bit smug.

Day 22: Calzadilla de la Cuerza

Many consider suffering to be part of a pilgrimage. I had hoped to avoid this component by doing my homework: training for months, buying the right shoes, the right socks, etc. It would appear the Camino has other plans for me. The agony caused by my feet is humbling. At the end of the past few days, my body has been so saturated with adrenaline released in response to the pain, that I feel weird and am not hungry.

I wake up this morning praying that my feet have healed overnight. I *have* to feel better eventually, right? This can't dog me through my entire Camino, can it? As I tentatively touch them to the ground at 6 a.m., my hopes are dashed. Kevin in the bunk above me hears me stir. When I return from the bathroom with some washing I had hung out last night, he's fully dressed and rolling up his sleeping bag. He is visibly antsy and ready to leave. Clearly *his* leg is feeling better, and I've barely begun the preparations for my feet. I turn on my phone, and it *tings* telling me I have a new text.

"Call me ASAP!" From Serena, the woman running my massage school.

Damn! She is not a drama queen, this can't be good. Kevin sighs audibly.

"Go on ahead, I'm gonna be a little while, I'll text you later," I tell him cheerily. I want him to know we're in a guilt-free zone.

We had not discussed walking together today, I don't think either of us wants that commitment. With a smile of relief, he leaves before I can change my mind. Frankly, I'm relieved too, I need some quiet time and once again notice that I really enjoy walking by myself and setting my own pace.

I quickly dress and throw everything in my pack, then walk across the street to an open bar. I take a seat on a stool and call Serena.

"Oh, thank God you called, I'm so sorry to bother you, I just don't know what to do," she blurts.

The waiter behind the bar brings me my café con leche and a croissant. I smile at him and push a couple of euros his way. Since there is obviously something I have to deal with, I will indulge in a luxury this morning, eating breakfast before walking.

"Ok, let me have it," I say, taking a sip of coffee. *Oh, damn that coffee is good!* I'm half listening, half feeling how good it is to be here in Spain on this beautiful morning, in this sleepy little village, miles away from all my day-to-day problems.

"A detective came in yesterday."

"What?!" – Now she has my full attention

"No way to sugar coat this, turns out one of our students is a porn star."

"NO! Seriously?" I say with a little laugh. "Is that – illegal?" This is so out of my wheelhouse, I'm not even sure what the right questions are to ask.

"No, but it seems she may have gotten herself into some other trouble, and now she's on their radar."

This indeed is big news! But at least the school didn't burn down. When she tells me which student it is, I nearly fall off my stool. Well, I'll be damned! That little Miss Oh-So-Sweet-and-Innocent is the last one I would have expected.

The initial shock over, I listen to the big drama unfolding at my School. I take a sip of coffee and a bite of croissant and notice how detached I am from it all. Yes, I'm shocked, and it's an interesting, if disturbing, story. But I don't feel compelled to get involved. I tell Serena to suspend the student in question temporarily. I don't really know what else to do. Especially after I find out that a client who I know is a high-ranking, vice cop has requested her for a massage on Friday. Do they think she's turning tricks in my treatment rooms? Is she? This is all way too complicated for me to figure out here on the other side of the globe. I'll deal with it when I get home. I feel as if

I'm on a time-out from my life. For now, everything is on hold. Besides, I'm done with my breakfast and I want to get going.

I set off on wobbly feet supported heavily by my poles, down the narrow, cobblestone streets. I glance in the darkened window of a closed farmacia, looking at the small array of offered goods, with unintelligible Spanish names. I love the foreignness of it all. The way it confirms my distance from life as I know it. The relief at feeling responsibility for just myself for a little while is palpable. I laugh and shake my head, "a porn star?" I murmur. How quickly one sentence can re-evaluate all the assumptions and stories you've made up about a person's life. Do we ever really know anybody?

The wildflowers here in Spring are a visual feast for the eyes. As I leave the outskirts of town, they line the roadways and are interspersed in the vast fields of wheat and yellow rapeseed. May is truly a gorgeous time on the Camino. I frequently think of Cooper and his love of flowers. I take pictures and send him at least one-a-day.

I reflect on the simplicity of walking solo and take a deep sigh, grateful for the silence and my own company. It's very challenging to plan on traveling with anyone here. Too many moving parts. Too many opportunities for a potential drama to bring the outside world in, as personalities clash. Deep in contemplation this morning, I see once again that the Camino is like a microcosm of life, a condensed version, people coming and going from our lives – Not a bad thing, just the way it is.

The weather is perfect for walking today; temperate and overcast. The first few miles are on a tranquil country road, the only locomotion I see is a horse and carriage. *Wait – what?* Those are pilgrims in the carriage. It dawns on me that he's offering transportation to the next village, a long seventeen kilometers away with nothing in between. It is the longest stretch on the Camino with no services.

As my feet complain to me, I'm very tempted to yell out, "YOO HOO" and wave frantically enough to embarrass my children. But then I notice the pilgrims in the cart are all laughing and passing around a bottle of wine. The party atmosphere doesn't feel right to me for some reason, not here on the trail itself. Then I observe a man in the back silently studying his boots, trying to look invisible, hoping no one will recognize him. I realize that would be me, so I just watch longingly as that ship sails. On the Camino, it seems that even when we are injured and have no other recourse, the self-induced shame of accepting a ride is universal. We came here to walk after all.

Soon we are on a more typical, wide, gravel path that appears to go on forever, and the sun has graced us with her presence. My feet intermittently feel better, I take advantage of those times to walk faster. But inevitably, with this much walking and no breaks, the pain returns. At some point, I pass the official halfway mark of the Camino Frances. I'm surprised there's not a flashing sign that says *Congratulations! You're halfway! You've got this!* Or at the very least a stone marker.

In the sea of farmer's fields, someone has mercifully placed two picnic tables and benches. I sit, loosen my boots and put my feet up. A beautiful, blonde Canadian woman sitting across from me smiles understandingly, "you're hurting eh?" She says with a small nod. My tough facade crumbles as I burst into tears. I hadn't realized they were so close to the surface. She launches into well-meaning advice, filled with time-worn platitudes, "the Camino provides, you're never alone on the Camino," yadda, yadda, which I find a little condescending. All I really needed was her presence and her ear. But I take advantage of the opportunity to vent my frustrations over the pain and let myself have a good cry. At the end of the day, I'm grateful for her kindness.

As I continue on, the inevitable happens, I need to find a pitstop. My bladder is about to burst. There are pilgrims everywhere, and with kilometer after kilometer of grain fields with no trees, there's absolutely no privacy. When suddenly I spot, *The Facility*. I had heard of this place. Urban legend has it that it's best avoided. But I'm desperate. Can it really be that bad? In my experience, whenever you have to ask yourself that question, you're usually in trouble.

I'm looking for a structure with four walls, but all I see is something resembling a duck blind. I look around the wall of straw and stop! An unholy mess, like every nasty diaper I've ever changed all in one place, greets me. At first, I recoil – nope, can't do it – but now my body is saying *hey, you promised.* I think a prayer is in order here. *Dear God, please don't let me contract some unmentionable disease, just because I can no longer control my bladder.* I ever so carefully pick my way back, just far enough to be out of sight, and am careful to touch absolutely nothing. I fairly levitate over the ground and set a world record for Speed Peeing.

Calzadilla de la Cuerza is a tiny village in the middle of nowhere, quite literally. It's surrounded by fields as far as the eye can see. It's only commerce: two albergues, a bar, a hotel with a restaurant that serves the only pilgrim meal in town, and a miniature tienda. This smaller-than-a-child's-bedroom store, replicated in many villages, offers minuscule choices of food and toiletries. It brings home the gluttony of alternatives available to Americans, promoting obscene commercialism.

Kevin greets me with his generous smile. Bless his heart, he's saved me a bottom bunk under his. A bonus, the Dutch woman Yoko who I keep running into is across from me. I wasn't even sure Kevin would be here. He usually likes to cover double this mileage in a day. But he doesn't want to push his leg too much just yet, and this was a long 17 kilometers.

"Go check out the bathroom," he says with the smile people get when they can't wait for you to open that special Christmas present.

Okaaay, this ought to be good, but it's a brand-new building, how bad can it be? I walk into a large room, freshly painted white. Four sinks are on the left, three bathroom-stalls on the right, five shower-stalls straight ahead. I later discover, that after installing four of the shower-door handles backwards, the clever handyman finally realized his mistake, but didn't bother to fix it. Hence only one of the shower doors lock. My eyes pan back still looking to find the source of Kevin's amusement, and I do a double take. What are two urinals doing out in the open, with not even little walls to separate them? I had missed them the first time as they look like the sinks they are right next to, the ones I will be brushing my teeth in later. I snort and bark a laugh. I walk out to see Kevin smirking and looking at me expectantly, we shake our heads and burst out laughing.

What on earth were they thinking, surely no one will use them? Do not underestimate the immodest European man. In a 24-hour period, I will interrupt two different men unashamedly holding their Johnsons, relieving themselves. I will quickly avert my eyes, attempt to stifle a laugh and backpedal to give them some privacy they clearly don't care about.

Having eaten only a croissant, coffee and orange juice for breakfast, my first order of business is a big bocadillo – I eat the whole thing. One of the best parts of this trip is the vast quantity of calories I burn every day. I lose weight as I consume food I would never normally consider eating, like a big, fat sandwich. And many days I experience true hunger, the sort that actually hurts. In a perverse way, it kind of feels good. It certainly makes food taste better. In our society, I think it is quite common for us to eat well before we reach that point. A hint of hunger sends us immediately to the refrigerator. Sometimes eating just because it's the right time of day, not even waiting for hunger to arise.

Kevin and I lounge away the afternoon in the shaded backyard, alternating chatting with silence. A sense of deep peace permeates everything.

Day 23: Zero day – Calzadilla de la Cuerza

My strong will is once again up against my common sense this morning. I make the difficult decision to take a day or two, off. It's kind of a no-brainer, as I can barely walk to the bathroom. What the heck is going on with my feet?

I awake to the crinkle of plastic, the hushed whispers between friends, a morning cough. It's hard to watch my fellow pilgrims get up early, the excitement of the unknown which accompanies all new days here, palpable in their preparations. With a smile and a nod first Kevin, then Yoko leave. With very little provocation I could cry. I'm wide awake and almost afraid of the prospect of a whole day with nothing to do, in a place where there truly is nothing to do.

So, do I take a taxi to the next big town 21 kilometers away, or just stay put in this quiet, one-horse village? I'm unable to make a decision about moving on so I stay put. Due to the pitiful state of my feet, the hospitalero agrees to bend the general rule at albergues on the Camino, that pilgrims can only stay one night. This lovely, inexpensive albergue, run by kind-hearted people is exactly where I need to be today. Besides, I might get to play reluctant voyeur again in the bathroom.

I talk to my youngest daughter Chelsea, in Hawaii, the only one still up due to time differences, and wise beyond her years. It is just what the doctor ordered to quell my frustration at being held back yet again. Every time I do short or zero days, any new friends I have made, move ahead of me and it feels like I'm starting all over. My initial fear of doing this on my own is replaying itself, over and over again, like an old vinyl record stuck in a groove. At home, it's easy to spend as much time as I like on my own because friends are just a phone call away. Here, once your friends move on, they're gone, sometimes for good. This leaves me feeling a little panicky.

They kick me out of the dorm room for a few hours for clean-up. I find a quiet spot, away from the depressing hustle and bustle of people leaving, to meditate. I stare at a donkey tied up a few feet from me, peacefully eating grass. He is carrying a large pack for his pilgrim – there are many ways to do the Camino. I listen to Amazing Grace and have a deep cry – for what I'm not sure. But afterwards, I'm feeling the Grace in my life again. How dare I feel sorry for myself in the middle of this adventure and the opportunity of a lifetime.

I see that I frequently compare myself to others. How far and fast are they walking? Do I measure up? Am I failing at this? The more I do this, the more my feet hurt. The pat response to all problems here: *You have to walk your own Camino*, can feel patronizing and irritating when you're up against your own expectations and competitive spirit. Saying, "*bugger off!*" to someone who's trying to be kind isn't a very pilgrimy thing to do, so of course, I hold my tongue. I do however promise myself not to advise anyone to, *walk your own Camino*, again.

So today I STOP, and I realize I'm hungry for this opportunity to do absolutely nothing. I bask in the peace and ease of life, in this tiny town. Like a fly on the wall, I watch their routines and rhythm. Part of me aches for this level of simplicity, another part knows I would be bored in two days, probably one.

I have been so caught up in *living in the moment*. My only job each day is to get up and walk. I realize, it's time to do the math. I just passed the halfway point, and I'm more than halfway through my allotted time here. Ruh-roh, this can't be good. I quickly see there's no way I can make it to Santiago in time for my return flight home at the rate I'm going. I am going to be forced to take some public transportation. Thank goodness my taxi cherry has been popped already, so I don't need to mourn that. I spend several hours poring over the maps and distances, trying to decide which is the best part to miss if I have to. I make no definite plan, but I now have a general idea and will trust my intuition as to when and where to fast forward.

Laying on my bunk, waking from an afternoon nap, my eyes focus on a familiar face across the room. I never thought I would be so glad to see Fred the Curmudgeon, from a few days ago. I ask him if he'd like to go sit in the bar for a snack, by now I'm hungry for company. He reluctantly agrees. I can't figure out why he seems so uncomfortable. Then he starts talking about his wife – OMG, he thinks I'm hitting on him. I stifle a smile, glad that he's one of the *good guys*, it makes me like him even more.

We eat lunch as we watch a spectacular storm roll through the plains. Loud thunder accompanies brilliant lightening-bolts and torrents of rain. It's powerful enough to warrant concern for any pilgrims caught out in it. One who comes in later to tell about it, said he crouched on the ground covered by his poncho, to make himself as small as possible so as not to attract the lightning. Imagine, everything is so flat from horizon to horizon, that a standing person becomes a magnet for lightning. God, I love this place! I'm also glad to know now what to do should I ever be caught in a storm here. Fred and I consider walking together tomorrow if my feet are up to it. But they won't be – sigh!

Day 24: Bercianos del Real Camino

There are so many ways to do the Camino. Yesterday brought two men with donkeys packing all their camping gear. Today I watch as a young couple, load their toddler and baby into the sidecar of one of their bikes, oh my! My mind can't even process how different their Camino is from mine. While I'm reveling in my newfound freedom, they're – well my God – they're doing this with two babies.

By default, my choice today is to take a taxi to my next destination, Bercianos del real Camino. My feet need another day of rest, but I have to keep moving forward. A call to my two daughters in Rochester, NY, helps me gain perspective and not indulge in self-pity for needing to take a taxi. My oldest, Michelle, exclaims, "Mom, we're so proud of you already for what you've accomplished, even if you had to quit now it would be amazing what you've done." Their encouragement helps me gain perspective and reconnect with my gratitude for having this incredible experience

The process of walking the Camino is stripping away so many layers of who I thought I was. I see that the gift for me of walking on my own is to find myself without influence from others. Who am I when I don't have to worry about what others think? As I drop my normal routines, habits, and rules, even the newly made up rules about how I should walk the Camino, a layer of persona goes with them. I am seeing that the real fear of walking on my own was not a safety concern, but something much deeper and scarier. It has afforded me the luxury of time to dig deep and find what lies beneath all those layers. I am slowly becoming unglued from the illusion of my identity, leaving me a little wobbly and simultaneously excited.

While waiting for a taxi, I check Facebook and see a message from Douglas. It's a belated birthday greeting. I want to reply with, *"I love you too,"* and a smiley face, as in his sweet, convoluted, non-committal way that's what he is trying to say. But I guess our fear of potentially spoiling a wonderful friendship by opening that door again is mutual. So, I cop out also and thank him in *my* sweet, convoluted, non-committal way.

I sometimes wonder, when we have our life review after all is said and done on this planet, how many times will I think, *how would it have been if I'd done that differently?* Will I laugh at my inability to do something as simple as telling people how much they mean to me? This I realize is a recurring theme in my life. It's some complicated, self-preservation technique that I use to distance myself and be safe from rejection. Anyway, the word *Love* is like the word *Aloha*, it has many meanings, most derived from subtlety and context, difficult to discern and easily misconstrued in a text.

At 10:30 in the morning, my taxi pulls up in the one-lane alley in front of my private albergue for the night. A slim, beautiful woman about my age with short, reddish blond hair and a kind smile, is standing outside the closed door. Annie (pronounced Ahnee) is from Holland, but her English is excellent. We soon discover we are both in need of a day off to heal. Neither of us is sure what time the albergue opens, but we are not kept waiting long. I have a reservation in the room with bunk beds. But Annie without a reservation will have to take a double room by herself, if she wants to stay here, as it's the only one left. Without thinking, I say that I would love to share that room with her if she'd like. This will open up my bed for another pilgrim who might otherwise be turned away.

The private room is the size of a phone booth, with two twin beds pushed up against opposite walls and four feet of space between them. The purple curtains and floral bedspreads make it feel homey. I immediately go to the small, windowless bathroom to take a shower. I push a button on the wall to turn on the light, not realizing that the clock is now ticking. The environmentally conscious Europeans have all kinds of ways to limit waste. In the middle of my shower the light goes off, and since I cannot reach it from where I am, I must finish in a strange space, in total darkness.

The coziness of our room invites openness. Between showering and settling in, we become fast friends. Annie and I discover a deep soul connection over a sangria with lunch. We return to the albergue to do laundry in the outdoor sink. After hanging our wash, we get comfortable in lounge chairs in the healing garden area. We are in a timeless bubble as we while away the afternoon, sharing intimate stories of this life and dreams of another, shared lifetime. We shed tears of joy several times as we help each other reconnect to higher purposes for doing this walk, and to see how unimportant many of the details actually are.

We talk late into the night, which translates to 10pm. We marvel at synchronicity, and how our Angels have worked overtime to bring us together in this lifetime to reconnect. Any misgivings I had about taking this second day off are quickly silenced as I wonder at the perfection of this soul meeting. This is one of several friendships I make on this journey that feel fated.

Day 25: Reliegos

Annie and I sit across from each other at the long breakfast table in the kitchen, with four other pilgrims. The man next to me silently devours two pieces of toast with jam in large bites. He washes it down with loud slurps of coffee, eager to be on his way. Annie and I look at each other, out of words, we smile, her eyes brim with tears.

As we finish our breakfast and prepare to leave, our hostess who speaks only Spanish, offers us both big hugs and kisses. Less than 24 hours ago she witnessed our meeting as strangers willing to share a room and can now feel our special bond. I sense she even feels some pride that her lovely home played a part, which it did, and which I now attempt to convey to her. "Su casa – umm," I put my hands over my heart, "muy healing." My Spanglish is atrocious, but she appears to understand.

Annie's ready to rock and roll, and I need to take it slowly. But we're both excited to be back to walking again. After the intensity of yesterday, we each need some alone time to integrate, so the parting is bittersweet.

The path today follows a country road with zero traffic. Annie quickly leaves me in her rear-view mirror, so for an hour I see no person or car. My gratitude at being able to walk again, albeit slowly, overshadows my still sore feet. I feel as if I could do this every day, just walk for miles. I pass a random Templar cross on a small stone statue, apropos of nothing. The mystery that shrouds the Templars fascinates me.

A couple going in the opposite direction, out for a morning stroll with their dog, smile and "Buen Camino" me. At the next town, eight kilometers and two hours later (it seems like a miracle that I have walked this fast), I see Annie again. It feels a little awkward and anticlimactic after our dramatic, tearful goodbye. We enjoy a French toast delicacy for second breakfast, then she goes to look for a room and me for a taxi.

I climb in the back seat as the young, nice-looking Spanish driver loads my backpack in the trunk. As we pull out of town and pass some walking pilgrims, I slouch a little, ashamed of once again having to *cheat*. I explain to the driver who speaks limited English, the problem I'm having with my feet. I know he could care less, but I want him to know I'm not just being lazy. The Spanish take the Pilgrimage very seriously and appreciate it when we do too.

Reliegos is the home of the Elvis Bar, famous for its cool name, two stories of graffiti-covered outer walls and I'm not sure what else, because it's Monday – and it's closed.

Moving on.

I find an aesthetically beautiful, new albergue, with lots of dark wood, tall ceilings and a clean bathroom. Everything is spotless and posh looking. I decide to stop here even if the hospitalero is a little taciturn. The room I'm directed to is small, with three bunk beds. I plop my backpack on a bottom bunk. My only companions for the night are two young, pretty, blonde American girls who give me a cursory hello without looking up from their phones. Ten minutes later, one decides to call her mom on speaker phone as if she's alone in the room. From this I learn they're sisters, one is sick with a fever, and they're missing home.

I'm not tired and don't wish to eavesdrop any longer on their conversation, so I head out to explore a little. I discover a small bar with a sandwich board out front displaying various paellas in beautiful, glossy pictures. I'm not really hungry, but a little bored, and everyone raves about paella, although I've yet to have one that really knocks my socks off. So, I order the vegetarian version. I'm finally sick of chicken and looking forward to a dish with a variety of veggies. I take my *vino tinto* (red wine) outside to find a table and wait.

Well yuck, that wasn't very exciting. Mushy rice, with a few sorry looking, mushier vegetables that all taste the same, served in a small, black, cast-iron pan to try and fool you into thinking it's fresh. I suspect the color is compliments of yellow dye #2 rather than the delicate spice saffron. I should have known this would be a commercial paella that comes frozen and is reheated. I've seen similar posters with the same pictures in many villages. I realize this is one solution to the problem, of efficiently feeding the vast numbers of pilgrims flooding the trail these days.

I poke my head into the small bar in my albergue to check out my options for an evening meal, since lunch was disappointing. An old, local man sits at one of the half-dozen tables reading a newspaper. The general air of the place is cold and lonely. I discover they do offer dinner, but in spite of the perfectly lovely décor, it just doesn't feel very inviting. Like the Tin Man from the Wizard of Oz, this whole albergue feels like it's missing a heart.

This town and this venue make me yearn for companionship. I haven't seen a single person I know yet, and the few people I've run into seem to want to keep to themselves. The other option I considered was the municipal albergue. But I'd heard through the grapevine that it's large and packs people in like sardines, so I decided to pass on that experience tonight.

For dinner, I decide to do something different. This place has a small kitchen just off the courtyard, available to its patrons. I don't see too many other options available in this small town. I haven't cooked for myself in over three weeks, and while I didn't want to make this trip about food, I am getting a little tired of pilgrim dinners.

When we drove through the town earlier, I noticed a small, colorful, eye-catching *tienda*. It doesn't open until after siesta at 4 pm, and by then there is a queue of people outside the door, I'm assuming from the municipal albergue. I'm happy to stand in line, even if I don't talk to anyone, at least I am around others and their conversations. I never realized I could miss being around people so much. Even though I'm single and live on my own, surrounded by nature, in a secluded place, I have anything but a lonely life. My world is exciting, rich and busy, filled with students, clients, and loving family and friends. I crave alone time and have to carve it out for myself. I guess that's very different from having it forced on you.

I see that one of the deeper lessons the Camino is teaching me is about the profound need we all have for interaction with each other. A year after my return, I followed on Facebook the Camino journey of a young, deaf woman. What was absolutely heartbreaking was her deep loneliness, as she had trouble befriending people because of communication difficulties. How often do we shy away from people who are different from us? Naturally we will gravitate towards easy relationships with people we can relate to. But how much responsibility do we have as human beings to also try including the 'others' occasionally, those without family and friends?

This store is larger than a walk-in closet, but just barely. Only half a dozen people, single file, fit inside, hence the line outside. The fresh produce is a hot item. I scan my few options as I get closer to the front. In a store this size, you shop as you wait in line, picking up things as you pass them.

On the one hand, you have to be patient if you see something up ahead that's going fast. On the other, you have to make quick decisions as the line is moving rapidly. Nothing so far is tickling my culinary fancy though, and I'm not really that hungry. I didn't walk much today, and the paella is still sitting heavy in my stomach. I almost feel I'm here more for the social experience.

I decide to keep it simple and choose an apple, some cheese, and half a baguette. As I get to the counter to pay, I spy a basket with several eggs in it. I don't know if I have a pan to cook them in, but these are sitting out, so they must be hard boiled.

"Un egg por favor," I say, forgetting the darned Spanish word for egg. I point to it, and the girl smiles. I guess I get points for using as many Spanish words as I can remember. She picks one out and wraps it in plastic. She then scribbles the price of all my purchases on a piece of scrap paper, filled haphazardly with other similar numbers. I watch transfixed, as she adds them up by hand and tells me the total. I smile at this quaint throwback to simpler times before electronics, when we actually used our brain to do the math. Seriously though, does she not have a smartphone with a calculator app? Somedays I feel as if I've stepped into a time capsule and been whisked back several decades.

As I prepare my dinner in the kitchen, I crack the egg to peel it and raw egg slimes over the counter. I laugh as I clean up the mess. I had thought it was a little odd that the salesgirl covered it in plastic wrap, now I know why. Thank goodness I didn't decide to save it for lunch, that would have been a train wreck in my backpack. I sit in the lovely, sunny courtyard with my feet up enjoying my picnic. I take the opportunity to go over the Frances app on my phone to see where I will walk tomorrow.

I'm feeling a little off, so I exchange a text with Annie, and then another with Kevin. It helps me to feel connected. That night when I crawl into my sleeping bag on the bed, I find it's a box spring, not a mattress. – Really? What a silly place to skimp on money. I toss and turn all night on my bed of nails.

Day 26: Hospital de Órbigo

The plan today is to walk as far as I can, then taxi to my destination to make up for the two days I took off with my injury. This part of the Camino is still flat and monotonous. I think I made the right decision to make up the time here so that I don't miss the legendary mountains of Galicia.

I stop for coffee and to use the bathroom in the first village. I'm sitting down when the light goes out after one minute, dejavú! I'm too far from the sensor to turn it back on again. I laugh and shake my head.

The walk along the road today is boring, and the concrete is hard on my feet. 16 kilometers of this is enough. Finally, I come upon a small town on a moderately busy highway. In a bar just opening, the waiter interrupts his start-up routines to make me a café con leche and *Naranja juma* (fresh orange juice). He's polite, but I can tell he's not yet ready for business, the muffins are still covered in plastic wrap. I sit at one of the dozen tables, the only person in the place, feeling alone and awkward. I'm not sure why, as I came to terms with eating in restaurants by myself, decades ago.

My feet don't get the rest they need, as I stay only long enough to finish my drinks. I'm in a strange headspace this morning. Everything feels a little off. I continue along the busy road, passing other shops and decide a distraction would be nice. I seldom listen to music, but this funk calls for a change of pace.

I am rocking out to the Eagles, starting to feel a little better, when two men come up behind me and slow their pace to talk. By now I am familiar with a normal pilgrim communication, so this one sets off my bullshit alarm. Instead of asking where I started, how far I've walked, or where I'm going, they are flirting with me, and I'm not buying it. They look like locals, but their English is good, and they have half-filled backpacks. They walk in front of me after I let them know I'm not interested. I watch as they approach another solo woman pilgrim.

All this happens as we are walking out of town and into an isolated field – Great! The few pilgrims that were within shouting distance are soon far ahead of me, and there is no one visible behind me. The two men slow their pace and are once again attempting to talk to me. With a shake of my head and a nervous smile, I point to my earphones to let them know I'm not interested.

I stopped this morning at two ATM's and between them withdrew 500 euros. Now I'm wondering if somehow, they saw this, and have been tracking me since then. I'm cursing myself at my stupidity for taking out so much.

The two men walk on ahead, seeming un-phased. They stop for a drink of water. I give them a wide berth as I have to pass them. They leapfrog past me again, and I studiously ignore them, pretending to be engrossed in my music. They stop to put on sunscreen, a strange thing for two men with their dark complexions to do. My heart is pounding, as this time when I pass them I have to walk into a creepy tunnel that goes underneath a road. There is no one in front of, or behind me that I can see – crap! Although I do have a *Hallelujah Moment* as I spot a village a kilometer in the distance.

The effects of adrenaline are nothing short of miraculous. My fight or flight response kicks in. I momentarily forget my painful feet and will my legs to carry me as fast as they can. I'm just shy of running, my walking sticks like a metronome, tapping out the rhythm. I'm glad I took the rubber tips off yesterday. In my head, I imagine using the pointy metal tips to fight them off. As I hear my labored breathing falling into a rhythm with my steps, the Lamaze classes I took when I was pregnant, come to mind. I smile a little at the randomness of that thought, then decide to go with it. I start repeating *hout hoot*, my favorite breath control technique, that distracted me through three deliveries. Who knew that was going to come in handy again? I don't look over my shoulder until I get to the town of Arcahueja, and by then there is no sign of the men. They were definitely up to no good. I know I dodged a bullet.

Any thoughts I had of attempting to walk into Leon today are shattered. The mental and physical effort it took to cover these 17 kilometers, including this last mad dash, has left me completely drained. After an hour of waiting for a bus that never comes, I call for a taxi.

I have no real desire to see Leon, one of the four major cities on the Camino Frances. My happy place is in the countryside. So, I decide to just get it over with and take transportation to catch me up with where I need to be to make it to Santiago on time.

"Welcome, please make yourself at home," says the long-haired, quintessentially-hippy owner of the Albergue Verde in Hospital de Órbigo. I could lose myself in his sexy Spanish accent and smile. I draw a deep breath as I enter the living room, taking in the incense, a Buddha statue and other spiritual paraphernalia. Oh my God! This feels like home. I could cry.

I recognize and greet the three Australian pilgrims I last saw in Logroño. And Gudren, the German pilgrim I keep crossing paths with is also here. I give a silent sigh of relief. I feel that I am back in the flow of my extended Camino family.

I'm shown to a room with sturdy, comfortable, wooden bunk beds. The hospitalero leads me into the bathroom. Am I hallucinating? Is that really a spa shower with water coming from every direction? With my all-is-right-with-the-world smile, I return to my backpack to retrieve my towel and toiletries. Luxuriating in that shower will wash this day away.

The Australians have arranged for a professional massage therapist to come to the house, but there is only time for three appointments. I must have looked really pitiful when I arrived, because the man in the group gracefully gives up his time slot to me. Kindness on the Camino is contagious. We are all the recipients of it from the pious Spaniards, who believe deeply in the Pilgrimage and their service to it. Many of us then, in turn, look for opportunities to treat each other the same way. I am so profoundly grateful, I could once again cry.

The massage is followed by a yoga class from the albergue owner, who studied in India many years ago. But wait, there's more –

as the soup is ladled into our bowls at the evening meal, the five-person staff sings us a blessing accompanied by a guitar.

Everything here is done with so much love and joy. The vegetarian dinner cooked with veggies from the organic garden is the icing on the cake. I wouldn't be surprised to wake up sucking my thumb after this much nurturing.

What a crazy study in opposites today has been. Love and fear juxtaposed against each other. I don't know what to think of it, so I don't. I sleep like a baby.

Day 27: Astorga

What a difference a day makes! Between the massage, yoga and a great night's sleep in a healing environment, my feet are feeling great.

Walking with no limp, I reluctantly leave my little taste of home and set off on a glorious morning. The rising sun creates a pastel palette in the sky. Last night, our hospitalero had called us all out to the porch to point out a shortcut through his backyard. It leads to a path through some fields, to bypass having to walk on roads this morning.

The first little hamlet, Opción, is only a kilometer away. The early morning silence is broken only by the soft padding of my boots on the cobblestones. The houses and streets are a study in beige, making the village feel clean and well cared for. Splashes of color from spring flowers explode from window boxes, a perfect complement to the neutral background. Somebody has strung small flags representing different countries across the street, welcoming people from all around the world. There was no coffee at my *green* albergue this morning, so I join a dozen other pilgrims in the only open bar, for a café con leche and a croissant. I sit by myself but am comforted by the conversations in several languages surrounding me.

Now that my feet are not hurting, I am once again able to bask in the extreme beauty of this walk: The gentle curves of the rolling hills, a wide, sandy, easy-on-the-feet path and the morning sun, a golden glow in a clear blue sky, softening the morning chill. All promise an extraordinarily beautiful hike today. I feel a spontaneous rush of gratitude burst through my body and notice how surprisingly close to the surface all my emotions are, the good, the bad and the ugly.

It's only another two kilometers to the next stop, Villares de Órbigo. But I'm determined to take it slowly today, and rest frequently. Besides, the bar in this village is just too welcoming. The backpacks and poles set against the wall outside, are an invitation to join the pilgrim morning ritual of second breakfast.

Gudrun is standing at the counter ordering a croissant. She invites me to sit with her and another couple. The *naranja juma*, is cold and delicious, but I'm not really hungry for the tortilla I ordered. I take a few bites, and I'm suddenly anxious to be walking again. The unexpected pleasure of normal feeling feet leaves me excited for today's adventure.

I gather my backpack and poles and set off on my own. There is a steady uphill climb away from the village. The dairy farm on my left effuses the familiar odor of urine and cow dung. I have loved horses and been around barns my whole life, so I find the smell strangely comforting. A little further ahead I notice on the right, several little huts with baby calves in them.

A woman pilgrim with an unruly mop of long, curly, auburn hair glinting red in the sun, wearing rainbow-colored knee socks, stands by one of the calves deep in thought. She leaves before I get to her and strides ahead. I think little of it, as I am reluctant to interrupt the transcendental-experience I'm having this morning, called walking-with-no-pain.

For the next seven kilometers, I see very few pilgrims, which makes me feel a little uneasy. I thought this was the stretch that a woman was murdered on two years ago. It turns out it is the trail after Astorga. But that doesn't help me and my overactive imagination, in this moment.

The undulating hills and sparse woods make for perfect hiking terrain. Any fears I have are quickly absorbed by the panorama of my senses. A soft breeze brings with it the green smell of the grasses and trees, with a faint, dank, under-current of rotting vegetation. No cuckoos this morning, but other birds serenade me with their song. I smile big, as I feel Mother Nature's joy at the bounty of Spring.

I become aware that I am no longer alone on the path. The steps, crunching in the gravel behind me, are faster than mine. I look over my shoulder to see who it is. A tall man approaches with purposeful strides. He surprises me by asking in an Irish accent, "Are you Kevin's friend?" I can't remember anymore how he knew that, but I laugh for some time over the weird synchronicities that happen on the trail.

"We met in the very beginning, there were two other Irish lads with me then too." He says in his charming, lilting brogue. I remember Kevin telling me about this group, how they'd walked several 30+ kilometer days together. After that, the troubles with his leg started. Our conversation is quickly exhausted, and he moves out ahead.

Suddenly an oasis appears in the middle of nowhere. It is a small fruit stand in a spacious open area where other pilgrims are gathered. I don't know where they all came from, I felt as if I were hiking by myself today. Gudrun is sitting on a large, colorful, circular blanket. Next to her is the woman with the bright, rainbow socks I had seen looking at the calf earlier. Gudrun introduces me to Rene.

Her first words to me are, "Sit down, take your boots off, you need to let your feet breath." Her command brooks no argument, so I acquiesce without resistance, even though I have been lazy about removing my boots at rest stops in a futile effort to save time.

Rene's red mane of thick, curly hair shines in the sun. She looks to be a decade younger than me. We have the same sense of humor and are laughing and having fun like old friends within minutes.

Taxiing ahead yesterday was a brilliant idea. Several familiar pilgrims at the fruit stand greet me, and Kevin's last text said that he is considering overnighting in Astorga this evening. I am back on schedule and have caught up with my Camino family. Joy, joy, joy!

As I prepare to leave the oasis, Rene laces her boots up. In unspoken agreement, we head off together. I have to pick up my pace a little to keep up with her.

"So where did you start?" I ask. Although I am distracted, as I look down at the two tracks we are walking in that bisect a huge, flat, open area. They appear to have been made by vehicles. It looks like the access path for trucks maintaining the electrical wires hanging high above us. Who else would be driving out here in the middle of a field?

"In St. Jean this time. This is my second Camino," she says.

"Really? When did you do your last one?"

"Last year," with a laugh "I don't know what it is, but I can't seem to stop thinking about it."

"Oh, crap," I say, "I was hoping once I got this out of my system, I'd actually be able to think and talk about other things – you mean this doesn't go away?"

"Nope, not for me anyway," she chuckles. "I live in Europe now, Germany, and I see random signs for it all the time. The first time I went to look at the place I'm renting, I saw that it had a yellow scallop shell on a blue background attached to the wall. I could hardly believe I was going to live in a place that is right on the Camino in Germany."

I stumble on a rock sticking up but manage to catch myself with my poles – I'm not sure how much longer I will be able to match her pace.

"The Camino is in Germany?" I ask.

"Yes, it's called Jakobsweg there! But I didn't know that last year. I was walking my dog in the forest and stopped dead in my tracks when I saw a yellow arrow hand-painted on a tree. I thought I was seeing things. But then I ran into another, and another. It's crazy." The remembering causes a momentary lull in the conversation. Her voice becomes softer, conspiratorial, "It's like she's following me."

"She?" I asked glancing at her with raised eyebrows and a smirk – and I thought *I* was obsessed.

"Yes," Rene whispers, putting a finger to her lips, "ssshh, she's everywhere."

We both laugh, then stop simultaneously to say "Buen Camino," to three pilgrims as they pass us.

"Well that sounds like a sign to me," I say, picking up where we left off, "I'd say *She's* not done with you yet."

"Oh, I'm not sure how much I believe in all that woo-woo stuff, but yeah, it is pretty interesting," she says with a smile.

"So, what brought you here the first time?" I ask, curious.

"Oh boy, that's a long story. The short version is, I had been offered this great job in Germany, but my husband didn't want to move. I needed some time to sort that decision out."

"Yeah, I can see where that would be tough, although sometimes a little distance can be a good thing in a marriage."

"Well, it was a bit more complicated than that – our marriage is good – a story for another day," she says with a wry smile. I glance at her, but her eyes are unreadable behind her sunglasses.

"So, what about you?" She asks, turning for a moment to look at me.

"I have no idea why I'm doing this," I laugh. "Just that I needed and wanted to. I keep hoping for some magical flash of insight, an enlightened moment. But when I'm not caught up with the amazingness of this whole experience, I'm just thinking about how much further I have to walk today, and when will my freaking feet stop hurting? Perhaps I'm just doing this because it's here, and I can."

"Hallelujah!" I sing as Astorga appears in the distance. I explain my *Hallelujah moment*. She smiles and nods.

We come to a large, concrete cross, on a hill overlooking the suburbs of Astorga. As we approach, we see an old Spanish man, bent over a walking stick, taking a picture of three women pilgrims. He hands back their phone, then turns his attention to us as we get closer. "Christians? Christians?" He asks nodding his head, in heavily accented English.

Without thinking I blurt, "Yes." Where did that come from? I haven't considered myself a Christian in years. Spiritual yes, Christian no.

He quickly reverts to his native language, Spanish, and babbles something unintelligible waving his free hand around. We look at each other puzzled, what does he want? He has kind of a creepy vibe. I can't figure out if he wants money, or to show us his penis, an appendage some old men are inordinately proud of. The Camino can be a magnet for flashers, the perfect audience streaming by daily. He inches closer, and grabs Rene's arm, pulling her towards him.

"Pray to St. James for me," he says, thumping his chest with the hand holding the walking stick, which is starting to look to me like a weapon. This is not an unusual request for the locals to make, but I feel something's not right, especially his physical contact. I grab Rene's other arm and say, "Let's go!"

She is indecisive, she wants to give him the benefit of the doubt. I firmly re-state "Let's GO!" She nods quietly, shakes him off with some difficulty, and we walk away. She rubs where he grabbed her and says, "That hurt like fuck! What the hell did he want?"

"I have no idea, but I'm glad we were walking together, that would have creeped me out if I was alone."

"So where is your husband now if you're in Germany?" I ask as we descend the hill into the suburbs surrounding Astorga.

"He's holding down the fort in the States. I miss him – we talk every night."

I don't mean to pry, but I'm always in awe of people that not only manage to keep their marriage together, but who are still happy after so many years. How do they do that? I've begun calling relationships *The Final Frontier*. In spite of all the other success in my life, I can't seem to get that part right.

My feet are really starting to hurt, I know I have to slow down. Rene has been telling me she needs to walk faster, to try and find a dentist to fix a loose crown, before siesta. So, we exchange contact info, and she moves off ahead.

That was the most conversation I've had in several days. The emotional rollercoaster is exhilarating and exhausting. I'm swallowed up in a long, anonymous stream of pilgrims. We move as slowly as the large snails prevalent along the trail. Like them, we carry our survival needs on our back. *One last push,* I think to myself, as I look at the steep hill that stands between me and this beautiful city.

The narrow streets are as pretty as an old antique restored to perfection. With enough *ancient* to give authenticity, and clean and well-kept enough to be inviting. I pass a lovely little church and jump as the bells toll once. It is half past something – I'm too tired to care what – my relationship to time has become a nebulous thing. When it's morning – I get up and start walking. When I'm hungry – I stop to eat. When I'm tired – I stop to rest. The actual numbers on the clock, mean little to me anymore. Without time to constrain me, to tell me what and when I should be doing things, my place in the Universe feels a little wobbly, not as certain as it once was.

My musings are interrupted as I see a narrow, cobblestone street open up in front of me to a huge square. It is a hub of potential activity, with spoke-like streets going out in several directions. Each side of the square is lined with shops and bars and rows of identical tables and chairs with white umbrellas.

My eye is drawn to a hand, waving like an overzealous traffic cop, beckoning me over. The hand belongs to Güdrun. Smiling broadly, she sits at one of the tables with two other ladies. But I am too tired, and each step sends a shooting pain up into my right ankle. The earlier joy of walking pain-free, was unfortunately short lived. I'm afraid if I sit down I may not get up again. I shake my head and mouth, "sorry." I just need to find my albergue.

Going on instinct alone, I cross the square and enter another alley. This one has several high-end, boutique dress shops. I window gaze as I continue my forward motion. Around the next corner, I stare in wonder at another glorious Cathedral looming high in the sky. I would not be surprised to see Rapunzel leaning out of an upper turret, her hair grazing the sidewalk. This Gaudi masterpiece causes me to stop for a moment to drink in the beauty.

To my utter relief, I discover that my municipal, 95-bed, albergue is just across the street from it, down a quiet alley. The busy hospitalero smiles kindly, but doesn't make eye contact as he signs me in. There's a sign-up sheet on the desk for massages. I draw a sharp intake of breath, hardly daring to hope. I point to it and say "massagee?" trying to make the word sound Spanish. "Si, si," he says pushing the paper towards me. I add my name to a time slot one hour from now. Just let me get a shower first, this comes before food.

I follow behind him as he leads the way up the stairs. At the second floor he stops and motions to the communal bathrooms. "El baño," he says, then continues to the third floor. We cross a large room, and he takes me to a real bed, not a bunk. Under normal circumstances this would be cause for celebration, especially for nine euros. But I'm still thinking about the bathroom, one floor below me. A painful walk and a flight of stairs away, in the middle of the night.

The large, dormer style window above my bed is cranked open, the rooftop next door is only 20 feet away. I can see the moss on the tiles and the peaks of other roofs in the distance. I look down to the street below and see two pilgrims turning into the front door. Very cool, the hospitalero has given me a really lovely location. The bathroom situation is momentarily forgiven.

I sit down hard on the bed and remove my boots and socks. I prop my pillow against the wall, lean into it with a sigh as I pull out my phone. I have a text from Kevin.

Having a good day with my leg, so kept walking past Astorga. Maybe I'll see you tomorrow.

After replying to him, I close my eyes for 15 minutes. I then grab my toiletries and a change of clothes and head off to see what today's adventure in showering holds for me. A man is standing at the row of sinks in his *tidy whities* brushing his teeth, a woman is washing her hands. I go around the corner to use the bathroom first. A woman is leaving one of the stalls with a wad of toilet paper in her hands. I discover why when I see that there's none in the first three stalls. I make a mental note to bring a Kleenex down with me in the night, as I'm guessing all the toilets will be sans paper by then. In this albergue, I guess it's every woman for herself when it comes to TP.

Steam rises from one of the eight stalls in the shower room, and condensation drips down all the doors in the windowless area. The white on white decor is spotlessly clean, just wet. It reminds me of a locker room in a public swimming pool. The stall I choose has some standing water around the drain on the floor, and the walls and door have a sheen of condensation. Forget hanging the clothes I'm going to put on afterwards over the door, and there are no hooks. I'm finding this is a recurring problem in the larger albergues. I brought an S hook as recommended, but so far, all the doors have been too thick to hang it on. I remove my clothing while holding everything else, a feat worthy of an Americas Funniest Video. I then put the dirty clothes on the floor in the corner first, piling the rest carefully on top.

Following my wet-shower experience, I receive a divine leg and foot massage. I'm not really hungry yet, so I pull a few snacks from my bag and put my feet up in the courtyard downstairs, then blog today's adventures.

Ting! Text from Rene; *Got my tooth fixed, wanna meet for a glass of wine?*

My feet vote NO, but it's still early on a beautiful day and it sounds like she's right around the corner, so I set off in search of her. After more texts to try and figure out where the hell she is, it turns out that Rene is waiting for me in a bar off the large square I'd passed earlier, about a half a kilometer away. My feet are hurting so badly by the time I get there, I'm in tears.

As we sip our wine, Rene looks up and mutters, "oh shit, incoming!"

I look at her puzzled, "what?"

She sighs "Since Pamplona, I've been walking with one of my students. When she heard what I was doing, she asked if she could join me. She's intelligent, a serious student and really sweet, so I thought 'why not?' Let's just say the Camino is bringing out a different side of her."

I look up to see a 20-something, young woman, pretty and voluptuous. She exudes confidence in her knowledge of the effect she has on men. Two European men of about the same age follow her, like dogs in heat. As they near the table, she giggles over something one has just whispered in her ear. She flips her waist-length, blonde pony-tail over her shoulder. Oh, my! A real, live Barbie doll.

The air is ripe with pheromones as they pull over three chairs to join us. Rene and I exchange a glance. They order wine as Rene introduces me to Missy. I don't remember ever being introduced to the young men, but it doesn't matter, they are interested in only one thing. Although, a little later on, one realizes he's losing the race with Missy and sets his sights on Rene.

157

He leans in and murmurs in her ear, in what he hopes is a sexy voice, "I'd like to fuck you." However, his German accent makes the word *fauck,* sound more like a chicken announcing laying of an egg and is a little hard to take seriously.

Rene laughs out loud in his face, then just as quickly her smile is replaced with a look that would grow icicles on testicles. She looks him dead in the eye and with a little shake of her head says, "That's not going to happen!" Fortunately for him she's in a good mood, or I suspect he'd have been wearing her wine.

We drain our drinks as we tire of this. Rene says, "Let's go across the square to that little sports shop with the obscene, blowup backpack outside, and get you a pair of hiking sandals." An impulsive decision, I will not regret. The hormonal trio barely notices we've left.

It feels like my feet have been let out of jail. My new, pink, Keen sandals are exactly what I needed. I don't quite have a bounce in my step, but I've stopped crying.

Neither of us are very hungry, we just want a bowl of soup. So far, the soup in Spain has been excellent, so we set off with high expectations. Three restaurants later, we finally find a bar that has *sopá,* although it's not on the menu – that should have been our first clue. It's a nice place, so we order a glass of wine, eat some delicious homemade, grainy bread and wait 20 minutes for – wait for it – chicken noodle Cup-A-Soup. It couldn't be anything else, no one could replicate that salty, gastronomic disaster. We are too tired to do anything but laugh, eat some of the noodles and go to our respective beds. There have definitely been some adventures in ordering food.

Day 28: Rabanal del Camino

What a blessing to walk with almost no pain. Or at least that's what it feels like after what I've been enduring. I'm in love with my new sandals. When I leave the albergue in the dark, around 7 a.m., very few people are on the streets. Walking out with me is a European woman I met several days ago. We stop at the first bar, that is just opening its doors. We enjoy a café con leche together, then she walks off, her pace much faster than mine. Shortly afterwards, Frank, a grey-haired, amiable gentleman, falls into step with me. We walk and chat for at least an hour. He has done 13 Camino's, one with a hairline fracture in his tibia (discovered when he returned home). He regales me with many stories.

Today's walk has been very flat so far, a welcome respite for my feet. The two-person path has either followed the side of a quiet road or gone through fields. At the first village, Frank and I stop for a café con leche. I have enjoyed his company but feel complete. I would like to walk on my own now and enjoy what's left of my golden morning hour with no talking. As I'm thinking how I will delicately word this, Rene's distinct colorful socks and hair come to the rescue. She strides towards us making a grand entrance. Frank takes the hint, finishes his coffee and leaves, never to be seen again.

The ease of being with Rene is reminiscent of a decades-long friendship. The Camino seems to time-warp relationships into fast forward. We agree to walk separately but make reservations at the same albergue in Rabanal del Camino for this evening. After she takes off, I text Kevin to let him know where I will be tonight.

The quiet morning is punctuated with the call of the cuckoo bird. The sun illuminates the path, and the wildflowers are in chaotic perfection. I hear complaints about the Camino being too crowded recently. While some days, like this morning, are a steady stream of people, I find most pilgrims to be respectful and quiet. The majority of us are here after all, for a similar purpose: To let Mother Nature support us as we take the time out of our busy schedules, to think, heal and examine our lives.

After another long flat stretch of walking, I see a church spire in the distance. The path is lined by waist-high rock walls leading into the next village. Rene is waiting at the first bar enjoying a *Naranja juma*. Regardless of walking speeds, the Camino uncannily aligns my journey with the people I'm meant to meet each day.

Now for the long, mostly flat, seven-kilometer stretch to my goal for the night. My, now painful feet, slowly traverse a pretty, but uneven path next to a pine forest. I come upon a young couple, her arm is linked in his. She is wearing very black, wrap-around sunglasses, which tells me she is either partially or completely, blind. They occasionally speak, quietly, in a foreign language and are in their own world. I marvel at the love and dedication between them, to attempt this arduous journey with her disability.

Turning a corner, I come upon a man dressed as a Templar knight, complete with a live falcon. He stands by a crudely fashioned rest area with a couple of rickety chairs. And then there's that! I gladly sit for a few moments, keeping a wary eye on the bird. He is standing on the ground, five feet away, his head is slightly cocked as he stares intently at me with one eye.

The next two kilometers entail some serious off-roading. I could have gone the blacktop route to my destination, but what fun would that be. I didn't realize how precarious this part of the trail is though. Large, uneven rocks going steeply uphill, followed by big muddy areas. I whack my foot on a rock, which given the state of my feet, takes my breath away. When I find my breath, I curse in a way that would make Rowena proud. The road is suddenly sounding much better. I find a path that, with a little navigating over a small stream, will take me to it.

A feeling of *coming home* hits me when I see the small crowd of pilgrims, sitting with their wine and beer, laughing and conversing. The Swiss Chalet look-a-like hostel/bar that greets me as I enter this village, feels like a step back in time. At my albergue a few streets over, a tiny, shaggy dog sits at the entrance, unimpressed with me. I briefly think of my two dogs at home, then just as quickly put them out of my mind. Any homesickness will be counterproductive to my journey here.

The busy, smiling hospitalero greets me and points to the bar to register for my bed. This two-story, U-shaped complex surrounds an open-air gathering area. Wash is hanging from lines strung in one corner and pilgrims are soaking their feet. Everything is bathed in the warm, late afternoon sun. Rene is here already, but the bunks in her room are all taken. I am given the first choice of bunk beds in another large room. I get a text, Kevin is on his way and wants to know where I'm staying so he can join me.

Meanwhile, a group of 12 Spanish teenagers show up looking for accommodations. It is clear from their animated conversation, and crisp, clean clothes, that they just started their journey. They're shown into my room – great! Ok, I will embrace this lively energy and hope they don't keep me up all night. I'm surrounded, resistance is futile.

Then Kevin comes to my rescue. He chooses a bottom bunk on the other side of the large room, next door to the bathroom, and suggests I move over to where he is – why didn't I think of that? I take a bottom bunk across from his, no one moves in above, or around us, we have a cozy little corner all to ourselves – perfect!

We return to the outdoor gathering area and pull up seats at Rene's table. I introduce her to Kevin. Missy shows up with only one of the men from yesterday. I introduce her to Kevin also, but she's distracted. She looks tired and worried and tells us she is having difficulties with her feet. She slumps down in a chair to show us some new blisters. It would appear the honeymoon is over for Barbie and Ken. He stands by like a lost puppy, and my impression is she's not sure how to shake him off.

Yesterday Rene had shared some of the drama going on between her and Missy. She hadn't counted on Missy, who is married, being the center of so much male attention. Missy's encouragement of it doesn't feel right to Rene and doesn't fit with her purpose for walking. Also, when they do walk together, Missy sticks to her like glue and chats incessantly. So, she has decided to put a little distance between them. I listened and nodded. I sense no connection with Missy, good or bad, so I will let Rene figure out her own karma with her.

Kevin and I decide to order an a la carte dinner at the albergue tonight and forego a pilgrim meal with Rene. It feels good to just have a large salad and a glass of wine. We retire early and quickly crawl deep into our sleeping bags as the room is cold. We turn on our sides facing each other.

"Do you have any pictures of your little house?" Kevin asks. We had talked earlier of my rather unique living situation. I live in a 146 square foot Tiny Home on wheels.

"Sure, let me look."

I scroll through the hundreds of disorganized photos on my phone. *When am I gonna get around to putting my pictures in albums?*

"Why don't you bring your phone over here," he says, moving over and patting his mattress.

Thinking nothing of it, I bunny hop over in my sleeping bag, and curl up into him, spooning in the narrow bunk bed. I wonder if he can hear the screeching of tires, as my mind and scrolling finger suddenly come to a rubber-burning halt. All my attention is focused on processing an onslaught of physical reactions, which catch me completely off guard. I am no longer in a hurry to find the photos. I feel Kevin's chest on my back and take in his body heat. I feel the comfort of his nearness, and wonder what it might be like to spend the night in the safety and warmth of a man's embrace once again. I want to close my eyes and soak in the surprising rush of emotions and feelings running through my body. It's been so long, I had almost forgotten how good this feels.

If I turn to look at Kevin in these close quarters, which I momentarily consider doing, it will be an invitation for him to kiss me. I feel a little wrench in my stomach, time slows to a standstill. I am scared and unprepared. Is that what he's waiting for? Or did this catch him off guard too? If I'm feeling this way, there must be something going on for him as well or is this just a natural bodily reaction after a 13-year dry spell. I'm clueless how to play this game anymore and not even sure if I want to.

I suspect this is a turning point of sorts in our relationship, a fork in the road. He extends an invitation, I accept, then clumsily falter, like an inexperienced teenager. Actually, as a teenager, I had more curiosity than fear and would have gone for the kiss. As an adult, all my experience has only managed to confuse me. My heart has been broken more times than I can remember, leaving me frozen in indecision.

I didn't ask, so I will never know what thoughts were running through his head at the time. In the way that men are from Mars and women are from Venus, his reaction, with a shrug may have been, "uhh, I just wanted to look at your pictures." – but I don't think so.

Ultimately, I'm afraid of rejection and potentially spoiling the wonderful friendship building between us. Also, new relationships are all consuming. I want my mind to be as free of clutter as possible, to glean the most from this pilgrimage. To be honest, this is all in hindsight. I can't claim to have had such clear, logical thoughts at the time. I was operating at the primal level of fear. Because when you're operating at the primal level of love, none of those excuses hold water.

So, I choose the safe route and with a deep breath and some difficulty, focus on my phone again. After 15 minutes of slowly looking through photos, we both appear to be procrastinating. There is no longer a reason for me to be there, without some kind of intimacy. Reluctantly, I not so gracefully, hobble off the bed and return to mine.

Kevin says wistfully, "I wish I were returning to Hawaii after this Camino. But I really need to go be with my dad before he dies. Besides, I've quit my job, all my stuff is in storage..." he trails off.

"I know, I know – I wish you were coming back to Hawaii too, but I understand that this is an important time for you and your dad." It sounds like we both wish we could have another chance at this in Hawaii, following our Caminos.

But I guess timing is everything. We allow the conversation to meander into a more neutral zone, before finally drifting off to sleep.

Day 29: Molinaseca

Kevin has set an ambitious schedule for himself today. Our flights leave on the same day, another interesting coincidence. He also wants to walk to Finisterre, 90 kilometers beyond Santiago. So, because of his time off for injuries, he needs to do some 30-kilometer days. He leaves as soon as he's ready, which is always before me with my foot taping regimen. I will not see him again until after I reach Santiago, but we will text each other daily with updates. Good news; no kissing, equals, no weirdness this morning.

Rene wants to walk by herself today, as do I. To ensure this (due to her ongoing drama with Missy) she leaves in the dark. We have agreed to meet later and share a room this evening.

The temperature is in the 50's as I leave at 7 a.m. But the line of pilgrims all soon shed layers of clothes, as we begin the seven-kilometer uphill hike to the Cruz de Ferro. Arguably the most important place on the Camino. The early morning sun is barely casting a shadow, as we walk up and around the side of the mountain. I am deep in thought as I contemplate the significance of this day. Pilgrims are told to bring a stone from home and symbolically lay their burdens down with the stone at the Cruz de Ferro (*Iron Cross*).

I explained the importance of this to my grandson Cooper. I asked him to choose a stone and hold it in his hands for a few minutes putting all his suffering and thoughts of cancer into it. Bless his heart, he sent two stones. One for himself and one for all the other little warriors fighting cancer. I cry softly to myself thinking of his compassion, as I fondle the stones that I have moved from my backpack to my coat pocket. As I walk, I concentrate on my intention to lay down the tremendous burden his little body is carrying. He is two years into a three-year protocol of daily chemotherapy treatment, including frequent spinal taps. His bravery is beyond anything I've ever endured.

The views become stunning the higher I climb. Foncabedon, looking remarkably like an Alpine-village, appears on my right on the side of the mountain. As I get closer, I can see that the village suffers from its remote location, the buildings are rundown, but many are still habitable.

I'm so excited when I see a sign for a homemade smoothie, that the owner of the bar becomes concerned. I laugh and try to explain that this is the first smoothie I've had in a month. Our communication barrier prevents him from understanding my outburst, he shrugs and walks off. My normal diet consists of a lot more fruits and vegetables and far fewer carbohydrates than I am eating here.

The path rises up ahead to my left and curls around the mountain. My sore feet are momentarily forgotten as I'm lost in contemplation on this beautiful trek. I call on Cooper's and my angels to energize this event.

The ten-foot tall cross stands in a clearing, atop a large pile of rocks and stones that others before me have left. A group of people is gathered around, laughing and taking pictures. Not the *angels singing, white light blazing down* experience I had hoped for. Magically, the party breaks up as I get there. I stand for a moment, tears well in my eyes. *Please Dear God,* I pray, (I considered praying to Mary, but decided to go straight to the *Head-Office* for this one) *let Cooper be fully healed and let this be his last experience with Cancer.* I follow a well-worn footpath to the top of the pile and place the rocks, with a posy of wildflowers.

Later in the afternoon, I return a message from my daughter Alicia, Cooper's mother. She opens with: "We got some really good news today."

"About Cooper?"

"Yes, you know how Cooper's treatment protocol was part of that study? And I was a little nervous because he's been getting less chemo and less spinal taps than kids on the *recommended* protocol? Well, Dr. Andolina called today to tell me that they've decided it will be the new protocol, nationwide, in the future, for children with leukemia. And they're even considering shortening it to two years instead of three. I feel like I can breathe again."

"Oh my God! That's awesome! So, you can stop worrying that he hasn't been getting enough chemo."

"Yes, it's such a relief."

"On another interesting note," I say, "do you remember those two rocks I had Cooper send me? He sent one for him and then he added one for all the other little warriors fighting cancer?"

"Yeah"

"Well, I placed them at the Cruz de Ferro just about an hour ago. It's the place on the Camino where you lay down your burdens, give them up to God." We are silent for a moment, my body is covered with goosebumps, we breathe in the synchronicity. Wow, that was fast!

Before the infamous downhill plunge from Cruz de Ferro, is a beautiful, long walk around the top of the mountain range. Somehow, I don't see a single soul during this time, giving me ample opportunity to integrate the morning. Then comes the descent from hell.

Thank god it is such a glorious day, I can't imagine doing this in the rain, fog or snow that frequently occurs at this elevation. I have to concentrate on each step. The loose shale on a downward slope is ankle turning, slip and slide territory, compounded by painful, unsteady feet. The camera on my iPhone stays tucked away in my hip bag after one quick shot. I will have to store these pictures in my head. It's interesting that following such a momentous event, I have to then surrender it and not think about it, to vigilantly keep myself from falling.

My nerves are shot, and my legs have turned to jelly, so I stop at the next village, El Acebo. I order a cold orange soda and a delicious bocadillo. A cold soda tastes so good after a long, hard walk. The dietary no-nos are accumulating.

As I sit outside in the sun, congratulating myself on another amazing day, I notice an unusually large number of boisterous pilgrims seated all around me. None of them have backpacks or a water bottle, and few of them even have *appropriate* footwear. Then I remember the big bus I'd passed at the top of the mountain. As I leave this town, I will see that same bus parked on a side road waiting for them. They call themselves *Camino lites*. While it's an entirely different experience than I am having, how wonderful for the Europeans that they are able to do this. I'd sure as hell take advantage of it if I lived here.

The yellow arrows take me on a small country road, an easy walk to the next village, Riego de Ambrós. But my feet are screaming, I'm done walking today. Luckily a tiny bakery is open in spite of it being siesta time. I choose a delicious-looking small tray of buns.

As I pay, I ask the woman, "un taxi por favor?"

My angels are with me, she says "si, si" with a smile. Then rattles off something to her husband, who makes a call for me.

Sitting on the wall outside to wait, I open the container and remove one of the sticky delicacies. It takes me back to a treat from my childhood in England, called dripping cakes. I look inside to catch the woman's eye and say, "delicioso" with a big smile. I'm not sure if that's Mexican, Italian or Spanish, but she seems to understand and nods, beaming. The taxi pulls up, I close the container and notice that it was a three-bun wait. I'm so relieved to not have to walk the remaining six kilometers to my destination for the evening.

Rene is already sitting at an outside table with a sangria when I arrive. What a welcome sight she is, sitting beside the river, the beautiful arches of a medieval bridge as a backdrop. She waves me over after I drop my bag at the private hostel next door. I break my routine as I decide to rest and have a drink before my shower today. I smile to myself at how fast we create new routines to replace the old.

Victoria, Colin, and Sophie from the English posse are sitting at the next table. I introduce them to Rene and invite them to join us. As they settle in the remaining chairs around the table, I observe Victoria taking in Rene's wild auburn hair, glimmering in the sun. Her glance moves down to her short hiking skirt and knee-high rainbow socks, as she sits casually slouched in her chair with crossed legs. Nobody represents their *back-home* self on the Camino. We all blend into this ageless pilgrim community, of *interesting* clothes, that are sometimes stained and smelly, muddy boots and au natural looks sans makeup and hair dryers.

Sounding uncannily like the Queen from Alice in Wonderland, emphasizing the 'you' by going up an octave, Victoria asks, "And what do *you* do?" With a lift of her eyebrow, she looks Rene directly in the eye.

Rene smiles, recognizing the unusual start to a pilgrim conversation. She sits up a little straighter and takes a slow sip of her drink.

Oooo, fun! Game on!

"I'm an English Professor at an American University in Europe. It's a temporary four-year assignment, my home base is in the States." She looks directly at Victoria as she delivers a perfect serve. I notice Rene becomes more articulate, her diction more clipped and precise when she's being challenged.

Victoria, momentarily taken aback, regroups, then comes back strongly with her own educational pedigree.

They volley back and forth as they discuss different, and obscure to me, literary works. I observe a subtle shift of power as Victoria now appears to be trying to impress Rene.

I turn my face to the sun as I take a sip of my sangría and laugh inwardly as they spar. Victoria has clearly met her match, she gives a little sniff and a nod of her head, as Rene passes some inscrutable test. Entertaining as this is, I can no longer put off taking a shower and getting into some clean clothes. I excuse myself. Rene and I splurged on a private room, with en-suite bathroom, at a lovely hostel tonight. It's funny to notice how excited I get about having a private bathroom and comfortable, twin beds with no one sleeping above me.

Tonight, we will eat at an Italian restaurant that Rene remembers having a really good meal at a couple of years ago. Unfortunately, their quality has declined. I have a terrible rendition of spaghetti, with a sauce that had to have come from a jar. It's pretty obvious that the food on the Camino has become industrialized, due to the vast number of pilgrims they need to feed.

Day 30: Villafranca del Bierzo

I hear Rene stirring in her twin bed next to mine, this time there is a nightstand separating them. I look over in time to see her leap out of bed, buck-naked, and strut unabashedly across the room into the bathroom.

'I love this woman!' I think to myself with a smile, as I close my eyes for a few more minutes of sleep.

"Well, look at you gorgeous," I hear coming from the bathroom.

I pick my head up again to see through the open door, that Rene is standing in front of the mirror addressing herself.

"And don't you look *fabulous* this morning," she adds, turning her head from one side to the other.

"You are so funny Rene," I say with a deep chuckle.

"My dad used to do this every day when he was shaving in the mirror," she says. "He'd say," She drops her voice low, and mimics a strong southern accent, *"well, aren't you a handsome devil Irv?"* - now it's become a habit of mine too." She laughs and closes the door. I think to myself, wouldn't it be a wonderful world if we all started each day addressing ourselves like this.

Missy stayed in another albergue last night, but she joined us for wine before dinner, and she and Rene have kissed and made up for now. So, the three of us decide to do something a little different today. We will walk a modest amount (16 kilometers), to save our feet for the upcoming O'Cebriero, touted as the tallest, steepest mountain climb on the Camino. Then we will visit a Templar castle, and taxi the remaining few kilometers to our hostel. We all have a schedule to keep, so some concessions have to be made.

I didn't go down without a fight though. I still have an inner conflict about taking a taxi. Rene looked at me with a *"really?"* look on her face. "You're a taxi whore now girl, you lost your taxi cherry a long time ago." Through the laughter, I understand – the purist in me has already been compromised.

There is a slight drizzle as we start our walk. Rene and Missy quickly move ahead of me. I find a rose lying on the sidewalk. As I put it behind my ear, the way we do with flowers in Hawaii, I feel a little jolt of homesickness. Fortunately, I'm distracted, as the drizzle becomes rain and I'm forced to dig through my bag for my poncho.

The English group descends on me as I'm trying to decide which way to go. There appears to be a choice, but I'm not sure. After a quick hello, four of them move out smartly ahead, with Sophie at the helm setting an impressive pace. She didn't even blink, as she decided which was the most efficient way. Not to be outdone, Victoria, Colin, and Kate-of-the-dirty-socks turn to follow her, struggling to keep up. They leave June behind without a backwards glance. June, still suffering from blisters, looks at me and smiles. I think she is relieved to have found a fellow gimp to walk with. She plans to take a bus from the next town to her destination.

I meet Rene and Missy at a sweet, little bar across the street from the Templar Castle. I get a smoothie, and two chocolate stuffed almond cookies with liquid chocolate to dip them in. Dear God! Only when I'm walking this kind of mileage every day, can I indulge in total decadence. The English cohort is in the same bar. June who knows of our plan to taxi after the Castle visit, asks if she can join us instead of figuring out a bus.

The Knights Templar Castle across the street beckons us, even though it is still raining steadily, and I know a lot of it will be outside. One of the primary functions of the Knights Templar was to protect pilgrims walking the Camino, amongst other more mysterious activities. They were all murdered by the King's decree on Friday the 13th in 1307. Purported to be a significant part of the Friday the 13th myth. Some of the Castle is intact. The views of the surrounding town from the turrets, are not only spectacular, but help me to understand why all these bloody villages are built on hills.

During our taxi ride to Villafranca del Bierzo, June talks non-stop, a running monologue, the way of people made nervous by

171

silence. When she gets out at the village before ours, the three of us look at each other with raised eyebrows and sigh.

The rain has stopped. Our hostel is modern and clean. However, the not-quite-hot-enough water makes the bath I've been longing for unsatisfying. As we drove through the town earlier, we noticed a park with a beautiful collection of roses in bloom. Following our disappointing baths, we decide to stroll through it, and yes, stop to smell the roses. At the far end of the park we happen upon, of all things, a dog show for Spanish Mastiffs, being held in spite of the rain. They are wet, muddy, bedraggled giants of dogs. Unlike the perfectly coiffed canines at the dog shows in Hawaii, I used to attend with my Cavalier King Charles Spaniel.

We explore the town and join the throng of people strolling the sidewalks. We look for a place to have an early dinner. It's that or wait until after siesta (7 p.m.) when none of us will feel like leaving the room again. We spot a bar with lots of patrons (usually a good sign) sitting at the outside tables. Feeling a little chilled, we opt to eat inside. The warmth hits us as we open the door and are directed to a table. I discover Caldo Gallego soup, a delicious specialty of the area filled with greens. I follow it with the best homemade paella I've had so far. What I lacked in kilometers today, I made up for with food.

Later as I sit on my bed blogging, Kevin texts me. Unlike us, he covered some serious ground today and is a couple of villages ahead. That man is on a mission.

Day 31: Las Herrerias

"What do you think of these blisters guys?" Missy is sitting on her bed with her foot turned upside down on her lap. Her resigned, sad face tells us what she thinks of them.

"We're walking up O'Cebriero tomorrow Missy," says Rene. "I'd give it a break today if I were you. If you walk on them now, they may burst and be too painful to walk on tomorrow – if it were me, I'd call a taxi."

And just like that, Rene and I are walking on our own again today. Missy's attitude towards me is dismissive, so I don't give her my two cents, which would have been the same as Rene's. She seems to me like a very bright and beautiful woman. Unfortunately, she is choosing on this journey to rely on her looks more than her brains to get by, with her giggly, hair-flipping Barbie act. For some reason, she's not letting me or for that matter, Rene, *in*. We all react in our own way to the survival mechanism that's triggered by the uncertainty of life on the Camino. This is her protection, and I just can't relate to it, not at this stage in my life anyway.

In an attempt to avoid the rain, forecast later in the morning, Rene and I decide to leave extra early today. We are up at 5, out the door by 5:45, sun up at 6:20. No other pilgrims are evident yet. The dim yellow street lights cast a mystical glow over the stone buildings of this ancient, medieval town. As we pass one street, the enveloping morning silence is broken by muted music and loud drunken voices. The Spanish are not averse to late night parties, sometimes going strong into the morning.

I stand for a moment on an arched Roman-bridge and look down into the churning water of a river that will accompany us for the rest of the day. As we leave the street lights and walk into the pre-dawn darkness, the shapes of the surrounding mountains hover above us. They loom overhead like helicopter parents, protective, close and just barely visible. Rene moves out in front so that we can each have some quiet time. It's Sunday and no traffic is on the road. My focus is drawn to the sound of rushing water, and the rising sun illuminating the majestic mountains. As a result of long, daily walking-meditations in nature, my mind is clear of normal concerns and spontaneously goes to a big life decision that I have been ruminating on.

I've been a massage therapist for 21 years, teaching it for 15 of those and I've owned my own massage school for the past 10. I've never wanted to stay in one place or do the same thing for too long. Typically before this, I changed jobs every 5 years. I've been feeling a little antsy for a while now. Even though I still have a passion for bodywork and especially teaching it, I wonder if there's something more. I'm 63, so the concept of security, even though I know it's an illusion, enters into the equation. But the freedom I feel here and now, to be myself and to have few responsibilities is absolutely intoxicating. Will it be possible to go back and pick up my old life, as if nothing has happened? As I walk, the river gurgling on one side of me, the sun coming up behind the glorious mountain on the other side, I ponder my options.

I catch up with Rene at the first bar. Inexplicably, there are pennies stuck in every nook and cranny of the ancient log walls. I puzzle over this oddity for a few minutes, then add one from my pouch as it seems the thing to do, but I have no idea of the significance. My mouth waters as I watch the man behind the bar pouring steaming milk into our cups, leaving a frothy trail. I can't tell you how good that first cup of coffee tastes when you walk seven kilometers for it.

We planned to travel separately today and meet up later. Rene walked the first week out of St-Jean-Pied-de-Port with her husband. Then she met up with Missy, who walks at the same speed, sticks like velcro, chats like a magpie, and gives her no space even though she asks for it. So, I understand her desire to walk alone. But we enjoy each other's company enough, that she ends up slowing down to my pace and we walk together for the rest of the day.

We pass through several small towns and stop at two of the churches. They are ancient, well cared for, with elaborate altar adornment. "Do you notice how front-and-center Mary is in all these churches?" I say quietly, as we stamp our credentials at the little self-serve desk at the entrance to the church and drop a euro in the donation box.

"What do you mean?"

"Well, if memory serves, in most American Catholic churches, Christ on the cross is usually the focal point, with Mary more off to the side."

"Huh, I hadn't noticed that," she looks back over her shoulder for a long moment at the altar of the church we are just leaving. "I'll have to pay attention and look for it. What do you think it means?"

"I read a great book before I came, in which she talks about the Goddess worshipping society that was here long before the Christians, and that this was a pilgrimage then also."

"I had no idea – huh, that's cool."

"Of course, the inquisition wiped out all the history – taking all the proof. But she conjectures in the book, that possibly they put Mary in a more conspicuous position so that the Spanish would be able to better relate and transition to a new religion."

There is a pregnant silence as we get back on the road, heading out of the village. "Ya know, I've been wondering," says Rene thoughtfully, "if my preoccupation with the Camino is trying to tell me to go back to the Church, re-embrace my religion."

"The Camino transcends religion, Rene," I tell her.

Her face lights up. "Oh my God, you're so right." She's also possibly relieved that I've given her a, *one-time-only, use-it-or-lose-it, get-out-of-jail-free card*. Religion/Spirituality is a very personal choice, that for many of us is a life-long journey of growth. I would never presume to tell anyone what to do, so I leave it there. We walk on in silence for a while.

With five kilometers to go, we stop for a sangría in Vega de Valcarce, a village with little gingerbread houses on the banks of the meandering river we have followed all day. The bar is dark inside compared to the brilliant sunshine we have been walking in. There are colorful pictures on all the walls. With only a few patrons, we are able to snag a table for four so that we have room for our packs and poles. The delicious homemade sangría is making me light-headed, as we prepare for the final segment of today's walk. As we start the inevitable uphill climb to our destination, my eyes are drawn to two huge dogs tending a large herd of sheep. They are the same color as the sheep and almost look like part of the herd as they hang out with them, watching for danger. These minor distractions help to take my mind off the pain in my feet, which slowly escalated to excruciating and has now moved into one of my ankles.

We join Missy in the bar at the private albergue we have chosen for the night. The tiny village of Herrerías, is at the foot of the mountain we will climb tomorrow. The three of us share a room and private bath again. To access our room, we have to go up a meandering staircase. It leads outside through a little garden, then inside through someone's living room, then up another flight of stairs to the third floor. It's fascinating to see the ingenious ways that villagers along the Camino, have added rooms to their residences to rent out to the, ever increasing, numbers of pilgrims.

Rene and I drop our bags in the room and take quick showers. Then we go back down to the bar for another glass of wine and some cheese and bread, while we wait for dinner. Missy whose taxi dropped her off hours ago, has been keeping our table warm with two other pilgrims.

I notice the Australian group of three who keep intersecting my walk. I last saw them at Albergue Verde a couple of nights ago. We smile and greet each other. The counterpoint to enjoying walking this on my own, has been the comfort and security derived from running into the same groups of people over and over again. Sometimes we share a drink or a meal, sometimes just a smile and hello. The camaraderie satisfies a primal need for company.

Day 32: Biduedo (via O'Cebriero)

Our task today; to climb the steepest peak of the Camino, eight kilometers straight up. It will be a day of humility and Grace.

5 a.m. comes dark and early. We have a big day ahead of us. By now we are efficient at packing up, tending to our feet and getting ready, so we are on the trail by 6. It is difficult to see the yellow arrows that guide us along the Camino in the absence of daylight. The one we almost miss, is painted on the ground next to a building. The daily search for yellow arrows in every conceivable place, is like a giant scavenger hunt and it adds a comforting element of safety to this grand adventure.

As we leave the village the road immediately starts to ascend, and our legs get a taste of what's ahead of them today. A group of six horses stand untethered, in the middle of the road under a dim, yellow streetlight. They barely acknowledge us as we weave our way through them. They wait wearily for their day to begin, which will entail taking some pilgrims to the top on horseback.

After a kilometer, we are directed off the road to the left onto a wide, muddy Camino path. The three of us agreed to walk together until daylight broke. So, it is now time for Rene and Missy to move out ahead.

I know that Rene really wants to be given some space on the trail and walk by herself. Missy has been told this point blank by Rene, but she chooses to ignore it and creates unnecessary stress between them. I let them sort this out and take my own alone time to enjoy this mind-blowing walk today.

Shortly the path makes a sharp turn to the left and ahead is a steep, very rocky ascent. While it is a challenging climb, the sheer beauty of the surrounding forest, in mystical semi-darkness, is breathtaking. I stop for a moment to listen to the shrill, cacophony of hundreds of birds, breaking the morning silence. Mother Nature's Symphony performing for my ears only.

The road goes up and up, the woods open to fields. The muddy, narrow track is cut into the side of the mountain, with pastures above and below. The low hanging clouds do not detract from the extraordinary views of tiny villages, nestled into the green of surrounding mountains. The promised rain is holding back thank goodness, but only barely. Most of the trail is a rocky, dirt road that could become a muddy stream in the blink of an eye. As it is, the walk entails carefully picking my way around mud puddles and large rocks.

After three kilometers, I hit the first village and can finally have a cup of coffee and an excuse to take a breather. There I see one of the horses, standing lethargically in the middle of the road. He knows the route well and has taken it upon himself to start the climb without a rider.

After a café con leche, it begins to rain just enough to warrant a poncho. The panoramic views make up for the burning in my legs, as I continue the beautiful, but relentless climb, in the on-and-off drizzle. It is humbling to be surrounded by other mountain ranges, as I make my way towards the top of another.

The next rest stop cannot really be called a village. It appears to be comprised of a bar, a dairy farm and a couple of houses. Here I pass my travel mates who are just leaving. I could join them but decide against it. The frequent breaks and going at my own slow pace are allowing me to really enjoy this walk today. So, the three of us plan to meet shortly in O'Cebriero.

The damp drizzle and wet clothes, caused by condensation forming on the inside of my poncho, leaves me feeling cold. The homey bar and another hot café con leche, are the perfect remedy. I order a tortilla out of habit but can only eat half. Some days my appetite seems to be waning, in spite of all this walking. I've come to the conclusion, that it's the adrenaline being released in response to the pain in my feet. The industrial strength ibuprofen takes the edge off, but pain is still my daily companion. In order to accomplish each day's activity, I am forced to go into a zen, meditative space, disassociated from my body. In some strange way, this is making for a beautiful experience.

As I leave the bar, I notice a large, ancient building to my right. Something about it piques my curiosity. Oh my gosh, it's a residence on top of a cow barn. I'm guessing it was built in a previous century to capitalize on the heat rising from the animals during the winter. Oh, the price to pay for cutting down on those heating bills, I can't imagine the smell that accompanies it.

And now I make a mistake. I hesitate at a fork in the road leaving this hamlet. I can take either a muddy trail to the top or continue on a paved road to the same destination. Another hiker walking by senses my indecision and tells me he's taking the paved road to avoid the mud. Hmmm, it's only another two kilometers, and not having to deal with mud sounds enticing – and so I end up missing a significant photo opportunity. A large concrete marker that announces you have entered Galicia, the final portion of the Camino. My disappointment is short-lived, however. I have survived the infamous climb to the top of O'Cebriero! And miraculously the weather has held out. The occasional sprinkles were no big deal, as I've heard that often this can be a cold, wet, snowy, not very pleasant trek.

I walk into a gift shop and see Missy. Behind her is Rene, who is simmering on the edge of a foul mood. Reaching the top of O'Cebriero was one of the high points of Rene's last Camino, and she is trying desperately to recreate the experience. But her memory of exactly where she had that bowl of soup, next to a fireplace with a friendly group of pilgrims, has let her down. Try as she might, nothing about this time is like the last. The bowl of soup we finally settle on, from an aloof bar owner, has to be sent back as it's spoiled. Well that didn't go well! We gather our paraphernalia and leave, without my much-needed rest.

I suspect Rene's darkening mood is partly because she can't seem to shake Missy. That combined with trying unsuccessfully to re-create the *first kiss* of her initial experience. I understand this to be a common pitfall for those who walk the Camino more than once.

Rene is intelligent and aware of this on a logical level, but momentarily her emotions are in control.

The church in the tiny village of O'Cebriero, is like the poor cousin to many of the ornate churches I have visited so far in Spain. Its lack of gold at the altar gives it a humble and unassuming air. What I notice on this day and in many small churches afterwards, however, is a distinct switch to an emphasis on Christ, and more of the masculine figures from the Bible, being front and center. This masculine/feminine balance has my attention for some reason. I feel there is a story being told here. A story hidden in plain sight. For me, the current one-dimensional history of the Camino, centered around the Catholic church, feels shallow and lacking. I have found history to seldom be that neat and clean.

I turn around after admiring an unusual statue of Mary and do a double-take. Hans, the German pilgrim from a few weeks ago, is walking towards me. I didn't expect to see him again. The puzzled look on his face tells me he thought the same thing. In lieu of a greeting, he says, "How did you get here?" He remembers how bad my feet were.

"I walked," I say with a grin. I know what he's getting at, but I'm going to make him work for it. He won't let it go.

"Did you...?" he makes the action of driving with his hands, as if it's too dirty to say out loud.

"Yes, my feet hurt too much, and I needed to make up time." I rationalize with a smile and a shrug, to hide my slight embarrassment.

Why do I feel like I owe him an explanation? Sigh! Guilt, shame, expectations – To my surprise, he doesn't say another word. He nods, looks down, gives me a quick hug and walks on as if I have disappointed him in some way.

After becoming aware of my own judgments and rules about this walk, it's easier to spot others. *Hmm, is this true? Or am I still feeling guilty for breaking one of my personal, cardinal rules, and he's just mirroring my own disappointment?*

As we prepare to go, Rene decides she needs some food. I take her cue and continue on ahead, but unfortunately for her, Missy remains clueless. I leave them to their dynamic.

The path winding around through the woods out of the village is extraordinarily beautiful. As it opens on to a wide, gravel road leading downhill, my reverie is broken when I see two women in the distance walking towards me – with clipboards. I am suddenly aware that I have not seen another pilgrim in a while. I read some time ago that when traveling alone, you have a heightened awareness when it comes to potentially dangerous situations. And so it is now. I pay attention to the prickling of the hairs on the back of my neck. One of the red flags I had heard about in the forums, is people walking on Camino trails going in the opposite direction. There are also plenty of warnings about the various scams the gypsies have, to try and separate you from your money. A common one is to use a clipboard pushed into your waist to distract you, while picking your pocket underneath it.

I walk to the other side of the wide road, and sure enough, one of them drifts that way also. I swallow my panic and walk back to the other side. She mirrors me. I remind myself that criminals, like any other predator, target weak victims. I square my shoulders as they get closer. She holds the clipboard out to me and starts to say something in Spanish. I say, a little too loudly and firmly, "NO thank you," and with a tight smile give them a wide berth. The woman bends her head dejectedly. Now that the feeling of danger has passed, I feel sorry for her. Poverty is such a difficult circumstance in any country. I'm sure it leads people to do things they would never ordinarily consider.

For two kilometers I walk downhill, towards a beautiful vista of farmlands below. I come upon a small dairy farm and a tiny church. The small windows in the church allow very little light to enter. I stamp my credencial, leave a euro in the small collection bowl next to the ink pad, and sit in a pew for five minutes praying for Cooper. I drink in the beauty of yet another of these lovely country churches.

I exit to the sound of cowbells. Big, friendly-looking cows are ushered out of the barn next door, right into my path. An old woman in a calf-length dress brandishes a large stick. I don't see her striking them with it, so I assume it's only a communication tool. One cow stops six feet away, turns and curiously blinks her big, brown eyes at me. I stop too as I spy her large horns, but the brief moment of eye-to-eye contact tells me this is a gentle giant.

I continue on to the tiny village of Hospital and stop at the only bar for lunch. Now I'm hungry. I order a tuna bocadillo and an orange soda. Oh dear, this orange soda thing is quickly becoming a habit.

My phone tings with a text. Hmm, strange, Missy doesn't usually text me. *"Where are you?"* She asks.

"In Hospital at a small bar eating lunch."

"Oh, good I'm not far away, I'll join you, Rene had to go back for her phone."

Turns out, they too had stopped at the church. A kilometer later, Rene realized she had accidentally left her only communication tool there while stamping her credencial. So, she is hightailing it back.

I'm half-way through my bocadillo when Rene finally joins us. Her gratitude for finding her phone has released her from the grip of her foul mood. She orders a soda and takes a big drink.

"FUCK! That was crazy!" she says with a laugh.

Now there's the Rene I've come to know and love.

"I practically ran back there, and this group of young kids was just leaving the church. I was so relieved to see my phone still sitting on the table. The thought of having to continue my Camino without it brought everything into perspective. What a waste of energy that was trying to re-create my last time up in O'Cebriero. Thank you, GOD!" She looks towards heaven, pumping her phone in the air. Instant karma at work on the Camino.

Eight kilometers remain to our evening reservation at a private albergue. My feet hurt badly, but I want to keep up with Rene and Missy. I have been walking by myself all day, and the company helps take my mind off the pain. But I am forced to walk faster than I want to.

The narrow, gravel footpath along the side of the road is relatively flat with only a few small rocks. Somehow, I still manage to stub the toe of my boot on the edge of a rock sticking up two inches. So ironic after what I have traversed all morning. Because I am holding my poles, with the straps around my wrists, I can't catch myself (which may have saved me from breaking my arm). I plunge straight forward like a tree being felled. My fall is broken, as my right knee and left elbow hit rocks. By some miracle, my face is spared as I instinctively roll to the side. The pain in my knee and elbow is temporarily excruciating. It knocks the wind out of me. In tandem with the pain, a panicky thought crosses my mind that I have just caused an injury severe enough to force me to quit my Camino.

Two women from Wisconsin are walking right behind us and witness the graceless fall. They join my two friends, and the four of them circle me with concern. They remove my backpack and boots and open my jacket. As I sit surrounded by love and all my gear, I finally surrender to a full nuclear meltdown.

Missy says petulantly, "I fell the other day on the way down from the Cruz de Ferro, I was all by myself, and it really hurt."

Rene sighs and says, "This is Angela's moment. Can we please support *her* right now?"

I am a sobbing, snotty mess. Blubbering and blowing my nose loudly on my kerchief. All the frustration of the pain I have been trying stoically to deal with for three weeks, is finally released. The women helping, intuitively know to just let me cry, and for two or three minutes they hold the space for me to let loose.

We don't at first notice a Korean male pilgrim heading our way. I can only imagine that he is approaching his worst nightmare. An emotional female gathering that he wants no part of, blocking the narrow path. We all look up at the last minute to see him, eyes downcast, carefully weave his way through the middle of our group. He steps over my legs as if I am a dead cat in the way and with no comment makes like the Roadrunner, with legs a blurry wheel. I stop crying. We all look at each other, speechless for a suspended moment in time. "Buen Camino," Rene yells after him. He does not look back as we all dissolve into hysterical laughter. That was just what the Dr. ordered to break the spell of my self-pity party.

I had chosen to wear my boots this morning because of the tough hike. But now I permanently retire them and leave them to swing by their laces on the back of my pack, for the remainder of the

Camino. This completes my *Bag Lady* ensemble. I slip on my much lighter sandals and take a deep sigh of relief.

After our brief, but intense connection, the ladies from Wisconsin take their leave. Camino Angels, never to be seen again. I wonder once more about the perfection of synchronicities on this sacred journey, too frequent to dismiss as mere coincidence.

Rene and Missy slow their pace a little to allow me to keep up easily. We are soon off the road and on another country path. This will be our last ascent for the day, and part of it is very rocky. The final kilometers take us through a dairy farm with an attached albergue. Its' outdoor lounging area butts up to the path we are on. I look longingly at the pilgrims sitting nearby, their bare feet up on chairs, nursing glasses of red wine and beer and cheering us on with *Buen Camino's*. The map says three more kilometers, I can do this.

We arrive at our Casa Rural in a small community of modern homes called Biduedo. We have a room with three twin beds and a private bathroom this evening. The large group dormitories in albergues were a fun part of the adventure in the beginning. But with a week to go, and my body breaking down, I'm grateful for travel companions that want to share rooms in private residences. The accommodations are more comfortable, and we have the added pleasure of each other's company.

There are no sights to explore this afternoon in this modern subdivision. We all take leisurely showers, even though the shower stall leaks all over the floor, leaving us with a soggy mess of towels to walk on. We complete the daily ritual of washing out our clothes in the sink. Then we rest on our beds for a few hours, fluctuating between napping and tapping on our phones. This has been a big day. Integration time is crucial.

We go next door for a pilgrim dinner at 8 pm. There is one other table of pilgrims, and one woman sitting alone, so we invite her to join us. She turns out to be lively and entertaining – a little too much fun, trying a little too hard. Another reminder that we all face our own demons, in our own way on this daunting journey.

Day 33: Sarria

My lesson for today, *don't squat to pee in a patch of stinging nettles* — oh yes, I did. Fortunately, it didn't hurt for long.

I slept in until 6:30 this morning. Last night I had arranged for a taxi at 8 a.m. to take me past the first, difficult, six-and-a-half kilometers to make this day more manageable. Rene and Missy will witness a cow giving birth on their way into Triacastela this morning. I don't think my feet could have handled it, even had I known. But that would have been an experience to treasure.

I try not to focus on how bad my feet really are, but at the end of the day, I can barely walk to the bathroom. Quitting was never an option. I have generous distances mapped out for my final week, so that I can make it to Santiago on time to catch my plane. So, it's frankly a little scary to know I have to let go of my last crutch, the occasional taxi. You have to walk with no public transportation for the final 100 kilometers if you want your Compostela at the end. That means after Sarria, no more taxis. Some people cheat on this rule, but no one checks, so it's an integrity thing.

The taxi driver does not like getting up early to pick me up, made clear the night before as we negotiated the time. I would have liked to start earlier, but 8 a.m. was a hard limit for her, and even that she wasn't happy about. She gets the final word, as she looks me in the eye and charges me 25 euros (close to the same in dollars) for less than a ten-minute ride. This is extortionate by Spanish standards. But I smile and pay up, grateful that because of her I only have nineteen kilometers to walk today.

I am dropped off in Triacastela and stumble slowly, like some giant praying mantis with my poles. I hope the ibuprofen will kick in soon and my feet will warm up and take me safely through the day. The middle of the main street is slightly wider than most. There is absolutely no traffic at this time of the morning, just the occasional pilgrim starting out for the day. The shape and color scheme of the buildings in this town reminds me vaguely of a Swiss village. Far from becoming indifferent, the appeal of these quaint villages continues to enchant me. They are all, *the same 'cept different,* each unique in their own way

Had I not already had a good breakfast while waiting for my taxi, I might have happened upon Rene and Missy. They set out before me this morning, and I find out later they stopped in Triacastela for something to eat. Rene will tell me of their humorous encounter with the English cohort. As she and Missy walked into the bar, she was greeted with, "Oooh, Miss Looong Socks," from a grinning Victoria. Rene's colorful knee socks had become her trademark on this trip.

Over a café con leche, Rene told them of my bad fall the previous day. Victoria, her nose slightly upturned, with a disapproving sniff, in her upper-crust English accent, had retorted, "she's not getting on very well, is she?" Our encounters with this group often leave Rene and me in stitches.

I choose to turn right on the street at the end of town and miss the ten-kilometer detour on the left to a monastery at Samos. I've heard much to-do about this Holy place, but at this point, enough is enough. My focus is on finishing. *Maybe next time* – did I seriously just think that?

A little bridge over a stream takes me to the countryside where I will hike the rest of the day. But first I find a hidden spot to pee. Ooo, that's odd! What's that prickly feeling on my bum? As I stand up, I notice the large patch of tall stinging nettles I had chosen to squat in. I roll my eyes and laugh. For the next 20 minutes, I have to resist the urge to scratch my butt, which might have provided amusement for the three pilgrims walking behind me.

The trail leads uphill through the woods, and soon turns into a small stream over rocks and mud. I zigzag from one wet, slippery stone to another, a practice in mindfulness as I strive to keep my feet and sandals dry. Later I run into two guys that have been passing me on and off all morning, they are very impressed that my Keen's are still so clean. I say, "I'm that good," with a wink and a smile.

It is not until I return home that I realize one of the greatest gifts from the Camino, was the opportunity to live for such an extended period-of-time, being anchored to the present moment. I am acutely aware of my surroundings without the myriad distractions of a *normal* life to keep my mind occupied. I note the subtle, distinct earthy smell of the wet forest. The delicate notes of birdsong are my only music. I am mindful of each footstep as I traverse the uneven terrain of a rocky, woodland path.

The uphill climb is long and intense. Although my feet are still in a lot of trouble, I'm impressed with how strong my body has become. I'm now able to traverse long, steep hills without needing a break. Each day is filled with self-congratulatory, mini-accomplishments. Opportunities to feel good about myself. As I exit the woods onto a small country road on level ground, the view opens up to a mountain across a valley.

In the last two days, the landscape has changed from flat open plains to wooded trails, with occasional breathtaking tunnels of green. With the exception of a few talkative groups, most pilgrims are respectful of the serenity and beauty that surrounds us. I walk a long time this morning before I run into civilization and the opportunity for a break. Finally, as I descend down one tunnel of green, it becomes a narrow, cobblestone path between ancient, stone residences. I spot the little fruit stand and smile in anticipation, as I see the joyful gathering of pilgrims. I find that I look forward to the socialization more than the actual food.

So many people return to Spain after a pilgrimage with hearts blown wide open, to serve future pilgrims in one capacity or another. And, so it is with the glowing proprietor of this humble establishment. He has made a sign, with a heart and an arrow in the middle that says: *Follow your heart, for it is your true compass.*

I soon discover though, that while on the Camino, it's important to follow yellow arrows first and your heart second. I smile and wave and confidently leave going the wrong way. Fortunately, I have only passed a couple of houses, when that little sign pops into my head, and I have a strong urge to have a picture of it. So, I turn back. Which is when I see the yellow arrow going off to the left. Whew!

That could have been a big disaster! I get a little visual of an angel winking at me, and I silently send up a big thank you.

I traverse more beautiful countryside before the trail parallels the road leading into Sarria. I am hot, tired, sweaty and trying not to think about how much my feet hurt, when in a singsong British accent, I hear a cheerful "Hello." A smiling Sophie, looking fresh as a fucking daisy, blows by me, a determined look on her face. Kate-of-the-dirty-socks follows shortly behind, huffing in her effort to keep up. She keeps her head down and says nothing.

Several minutes go by before Victoria, looking decidedly not so fresh, starts to pass me. She barely has the energy to converse, but can't help herself, wondering how I got ahead of her.

"Where did YOU start this morning?" she asks accusingly. Her voice breaking on *YOU* as she accents it by going up an octave.

I know she's fishing to see if I will cop to taking a taxi (which I find out later she already knew as she had spoken to Rene). But I just answer, "Biduedo" with a little grin. I'll let her do the math. With a tight smile that doesn't quite reach her eyes, she walks on.

Phil passes me next, head down, quietly struggling to keep up with his wife. By his rolling gait, I can tell his feet are hurting too. Not far behind, June brings up the rear, even her natural cheeriness is being challenged as we all approach the end of a long, hot day. We have a short conversation, as we've bonded a little more than the others due to our painful feet. But she feels the pull of her tribe and soon walks ahead.

After a full, peaceful day of walking in Nature, I start to check out of my body as I enter the large town of Sarria. I wish I'd been more diligent with Duolingo to learn Spanish. I attempt asking directions to my albergue, but even their hand signals appear to be in another language.

Oh, dear Mother-of-God, are you KIDDING me with these inclines? I stare up at the impossible grade of the one-lane road leading to my home for the evening. It is of course about a quarter of a mile up, at the very top. I take a deep breath. I'm tempted to cross myself even though I'm not Catholic. Maybe I'll cross myself anyway, I need all the help I can get right now.

Rene, Missy and I decided on a private apartment connected with Albergue Matias for this evening. I use the term *connected* loosely, as in reality it is one street over, 100 yards away and on the third floor. Their genuine warmth and hospitality, however, allows us to forgive them for not mentioning this minor detail when we placed the reservation.

Tonight, we enjoy real Italian food. So far, the pasta on the Camino has been reminiscent of Chef Boyardee. Barely edible for someone who was married to an Italian for eight years. My ex-husbands, off-the-boat grandmother, passed down her Sicilian, family recipe for *Sauce* to me, so that I could be a good wife and make it every Sunday.

The real bonus tonight though, is our cute little, Johnny Depp look-a-like waiter. We overhear him at the next table offering the dessert choices in a deep, unhurried, sexy Spanish/Italian accent, lingering over the vowels. "Panna Cotta, Torta Cioccolatino..." We giggle like teenagers, and when it's our turn after he goes through the list, Missy flips her blonde ponytail over her shoulder and says innocently, all eyes, "I'm sorry, could you repeat that please?" I snort through my nose and cough, as I try to stifle a laugh. I can't even look at Rene – we'd be on the floor. He sizes up the situation quickly, and with a knowing, sexy smile takes a deep breath and repeats himself for the easily amused Americans. Because we haven't embarrassed ourselves enough, I then ask if I can take his picture. This is a big step for someone who cares so deeply about what complete strangers in a country half way around the world, think of her. I've come a long way baby, in more ways than one. The Panna Cotta tasted as sexy as it sounded.

After dinner, we walk back to the apartment. Rene and Missy have taken the two beds in one room, and I have my own room. We take showers and relax. Rene sits on her bed, scrolling through her phone looking for accommodations for tomorrow night. She has gotten her period, is a little cranky and is done being subtle with Missy.

"I'm walking by myself tomorrow, and I'll be staying on my own," she announces loudly enough for me to hear in the next room.

"And I'm doing a Camino reset with you too girlie," she adds. I know she's referring to me.

I walk to their bedroom and stand in the doorway, "You're breaking up with me?" I say with a mock sad face. She knows there are no hard feelings. She has already told me her policy on this, and I'm way past my three-day expiration date. But I will miss her for sure.

Day 34: Vilacha

Today is the day of cow shit! The air is permeated with it. I step around it all over the trail. I pass one small, local dairy farm after another.

I leave before everyone else this morning. I have developed my own rhythm and strategy for this trip. I like to get to my destination no later than 1-2 pm so that I have time to rest, do wash and enjoy the new place. I know that I take longer than most pilgrims due to my feet and a slower natural pace, so I like to leave early in the morning.

I have come to love the way the street lamps cast their golden glow in the pre-dawn darkness. It's hard to describe the amount of pain I'm in first thing in the morning before the ibuprofen kicks in. So, focusing on details like the natural beauty surrounding me helps. I take small steps, leaning heavily on my poles, especially as I descend the very steep, cobblestone road leaving Sarria, knowing that a misstep will cost me dearly.

On the other side of the cross-street at the bottom of the precipitous grade, a concrete marker with a yellow arrow points me to a dirt lane leading off into the tall grass. Up ahead I see a railroad track that the path parallels for a while.

Only a couple of other early-birds have passed me, so I'm surprised to hear my name called out. I turn to see Hans (my German friend) beaming at me. The joy and surprise of re-discovering friends on the Camino is a delightful mystery. Hundreds of thousands of people each year, walk at their own pace, across an entire country – it never ceases to amaze me. He's surprised to see that I am still walking while in so much pain, he thought for sure I would have given up by now, but I brush it off, and we exchange stories. We part this time with a heartfelt hug, he apparently has forgiven me for taking a few taxis. I sense this is our completion. We will not see each other again.

The light this morning casts a golden glow over the forest, illuminating the gnarled trunks of centuries-old trees that line the path. I stop to examine one extraordinary giant that could house hobbits. Someone has capitalized on this, can't-miss photo opportunity, and placed a donativo fruit stand next to it. I purchase a banana, always happy to support the local economy, and continue on ever upwards.

The trail continues up and down, up and down, through woods, small villages and beautiful lanes. It must have rained last night as evidenced by little streams running down the path today. One treacherous descent with boulders, mud and a fast-flowing stream, takes all my attention, as a line of us carefully pick our way down. No one even cares when we're splashed by two cyclists that pass us on the right, in the worst part. That's how intent we are on not causing bodily injury to ourselves. At the bottom, not surprisingly, one cyclist has a flat tire. I take a deep breath, roll my shoulders away from my ears and congratulate myself on my miraculously, dry feet. I wonder what this stretch is like on a *bad* day?

With a smug smile, feeling quite pleased with my navigation of that last descent, I cross a one lane road and follow the yellow arrows through the woods. But my brow furrows, and I cock my head to the side – *is that bagpipes?* I check my phone to see if I have accidentally turned something on. I did download Il Divo's version of Amazing Grace the other day, could it be that? I look around for evidence of homes, but I'm in the middle of nowhere, fields on one side, forest the other, I haven't passed a house in a while. It's so faint at first, fading in and out, I genuinely wonder if I am now hearing things. But as I keep walking, it gets louder and undeniable. Finally, I turn a corner and there, standing in a small clearing on the side of the path is a man, dressed like Robin Hood, playing what I later find out to be Celtic pipes (similar, but smaller than Scottish bagpipes). You just never know what you will encounter here. For a few minutes, I join the small group gathered to listen to this little piece of magic in the woods. I drop a couple of euros in his music case then continue on with a full heart.

The rest of the walk today is uneventful, several kilometers on the side of a quiet road. Hot and exhausted, I finally get my *Hallelujah Moment* as I spot a small village just ahead. I pray this is Vilacha – I have been fooled before.

My albergue this evening, Casa Banderas, is run by a delightful South African man around my age, a dead ringer for Robin Williams. Walking the Camino changed his life many years ago, he returned to Spain to purchase this beautiful ancient home and turn it into a private albergue. I notice right away a twinkle in his eye and feel some chemistry between us.

A text from Rene tells me she's somehow still behind me today, but close by. She gets directions, and we agree to meet at *my place* for a drink. While I wait for her, I ask Gordon if he can make a Tinto Verano (red wine and sparkling lemon) for my friend and me. "I think so," he says as he quickly disappears into the house, after showing me where I can arrange two chairs in the shade in the garden. Rene arrives as Gordon emerges from the house with three drinks. Well, ok then! He pulls up another chair on the opposite side of my friend, the seats are now arranged in a semicircle, and he joins us. I really enjoy the expectancy of inclusivity on the Camino. Americans are sometimes so prissy about their privacy and boundaries. Why can't we all just be friends?

Rene is sitting upright, legs crossed, wearing a short hiking skirt with her colorful socks. She runs her fingers through her loose, shiny auburn hair and somehow manages to still look bright and ladylike after a long day of walking. I on the other hand, still have my hair in a messy ponytail, that stopped trying to look feminine hours ago. I'm wearing a weird concoction of clothes and am slumped tiredly in my chair, legs a little splayed, on the opposite side of her.

Rene is married, and not remotely interested in Gordon, but she can't help herself. She's naturally vibrant, flamboyant, flirtatious and way more social than I am. She and Gordon are involved in a lively conversation about books as I tiredly look on. She giggles and flips her hair over her shoulder, in the age-old, ultra-feminine gesture of vamps the world over, and it hits me like a brick. A little voice whispers clearly in my head, "You've lost your *game*!" Time stands momentarily still, their conversation like Charley Browns' parents, waa waa, waa, waa. I slowly sit up straight, casually take my scrunchy out and run my fingers through my hair bringing it forward over my ears and around my face. Gordon flicks a look at me without missing a beat in his conversation with Rene. I scrutinize my mentor further and cross my legs. When did this happen? When did I forget how to flirt, how to be feminine? Ok, so it's been 13 years since I've had any kind of romantic interest, but really? It's not like I'm looking for a relationship at this moment. But I would like to bring back parts of me, that like the string of a helium balloon slipping through a toddler's fingers, appear to be lost forever.

I begin to see that part of this trip, is about re-discovering something I didn't even realize was missing. I have to chuckle at the timing though, that I chose to do this while wearing the same two sets of clothes and no access to a blow dryer for 39 days. But I see that it's a state of mind and really has little to do with outward appearances.

As Rene leaves, she hugs me and whispers in my ear, "I think he likes you." I laugh and pretend to shrug it off. Women love the intrigue of romance, whether theirs or another's. Whatever it is, I'm going with it. The anonymity inherent in this trip allows for defenses to come down, that have been fortress-like for years.

After I've taken a shower, Gordon calls to me from downstairs, "Would you like another Tinto Verano, this one's on me?" I smile to myself. "Yes, thank you so much, I'll be right there."

We sit next to each other on a couch in the large, tastefully decorated common area, with a ceiling so high a helium balloon could easily be lost, but I'm hanging on tight this time. I have nowhere to go, nowhere to be, but right here, enjoying this lovely man's company. We are laughing and exchanging stories of our lives when we hear a loud, coarse "Hola!" at the open front door.

Gordon stops mid-sentence, his eyes wide, and puts his index finger to his lips telling me to be quiet. But to no avail — without waiting for a response, an overweight, middle-aged woman, with short dark hair and a barrel-shaped abdomen, opens the door and strides in. Uninvited, she plops herself down in a chair opposite the couch we are both sitting on. Briefly looking me up and down she smirks, then quickly dismisses me. The spell is broken. Gordon recovers himself and introduces me to his next-door neighbor Marci. I later learn that walking into each-others houses without invitation is one of the local customs. Also, that his freezer is filled with her homemade bread. Apparently, the neighborhood women are taking care of the lone single-man.

Gordon and Marci are communicating in rapid Spanish. I watch spellbound, as without missing a beat in their conversation, Marci begins picking her nose intently using the thumbs from both hands, in BOTH nostrils. When I mention this later, Gordon rolls his eyes as he tells me it is another habit of the local women.

A few minutes into their discussion, Marci flicks a look at me and asks Gordon a question with a smirk and a little raise of her eyebrow. He answers her in Spanish with a little laugh, then suddenly his smile drops, "Do you understand Spanish?" he asks me alarmed. I chuckle and say "no." I don't mention that I can read universal body language though.

After Marci leaves, Gordon excuses himself to make dinner for tonight. I remain on the couch with my phone and finish my blog for the day. A delicious aroma wafts from the kitchen. I peek in to ask if he would like any help and see that he's making spaghetti Bolognese. Hmm, I know I took all the restrictions off my food for this trip, but my culinary meat excursions have not gone beyond chicken yet – am I willing to do beef? He declines my offer of assistance, his obvious joy in the kitchen makes my decision for me. Right, tonight, beef it is!

The only other pilgrims staying the night are a delightful couple in their 60's, from Ireland. They started their pilgrimage today in Sarria, as evidenced by the excited sparkle in their eyes, unlike the tired acceptance in mine. Gordon joins us for dinner, and the conversation is lively and fun. The three of us get some confusing directions from him for the following morning, with advice to miss a treacherous portion of the trail on our way into Portomarin.

I have the communal room with six bunks to myself tonight. The gentle breeze coming through the open window is like a lover's fingers on my face. I drift into a peaceful sleep, brought on by the deep satisfaction of a good days walk and the delightful aftermath, with dreams of helium balloons defying gravity and returning to my grasp.

Day 35: Eirexe

Gordon provides a stellar breakfast by Camino standards, complete with a soft-boiled egg. Well worth the four euros and a little later start. I have truly enjoyed every moment of my time here. Gordon follows me to the door, where I claim my poles from the basket, and we say goodbye with a long, heartfelt hug.

"I wish you could stay longer," he murmurs in my ear. I give him a peck on the cheek, "me too," I say. Although truthfully, while I feel a connection with this beautiful man, at this moment I am full, and ready for a new day on the Camino. I sling my backpack over my shoulder and clasp it across my chest and around my waist.

The Galician mist, otherworldly and mysterious, shrouds everything for the first hour. I struggle to make sense of the confusing directions to the easier path leading to Portomarin. I'm so grateful when the Irish couple come up behind me and seem to have a firm grasp on where to go. I follow them down a steep, narrow, rocky, muddy lane wondering if we have gone the right way. But I have since seen pictures of the *other* path, and it makes this one look like a botanical garden stroll.

Like Camelot in the mist, I see Portomarin on the opposite side of the wide river I need to cross. As I make my way over the long concrete bridge, I stare in awe at the town that looms above a tall flight of, what looks to be about 100 stairs. Thank God I will not have to climb them, as the yellow arrows take you around the town if you don't need to sleep there. I cannot imagine climbing those stairs, as many do, after a long, exhausting day of walking. Good call on many levels staying in Vilacha last night.

I am soon ascending up through woods again. This morning the trail is dry and level, and the mood is quiet and solitary. At some point, my feet stop hurting. I realize this is the effects of ibuprofen kicking in, and not some miraculous healing, so I take the opportunity to pick up the pace and cover some ground before the pain returns. This morning it hits me that I only have a few more days of walking, so I do my best to clear my mind and be really present with the experience, to gather all that I can from it.

By 11 a.m., the sun has broken free from the clouds, I know that the rest of the walk today will be hot. After hours of walking, I finally see a sweet little bar that is buzzing with pilgrims. The decision to stop for lunch is a no-brainer. I see Missy with a group of young people, she appears to have found some friends her own age, good for her. She catches up with me as I leave, and we walk together for a short while. She is dealing with her own demons on this trip, as we all do. But like a pair of mismatched shoes, we can never seem to make a deep connection. She soon moves on ahead of me.

With only a few kilometers to go, I stop for ice cream. I ask my feet to carry me just a little further, as I sit in the shade for a few minutes, with the ice cream cooling me from the inside out. For the final leg of today's journey, I put in my earbuds to listen to music to take my mind off the pain and exhaustion.

Airexe, my destination for tonight, is barely more than a few houses and a bar. I am so grateful to have a private room reserved in one of the two albergues in town. As I check in, I watch the hospitalero regretfully turn away two older women, who look as hot, tired and spent as I am. They ask if she will allow them to sleep on the floor in their sleeping bags, but she sadly shakes her head "no." I can see it breaks her heart to have to do this. I feel a moment of guilt for my own good fortune. But I can see that my lesson learned in Zubiri on day-three, a lifetime ago, has served me well. These poor women will need to walk a few more kilometers to the next village and hope there is an opening there.

I get a glass of wine from the bar across the street to bring back to my albergue. The little table and chairs set up outside for hanging out, are more inviting than the inside of the bar. As I cross the country road, wine glass in hand, I cock my head and squint at a woman walking towards me. At first, she looks at me blankly probably wondering why I'm staring. Then recognition dawns on me.

She is the friend of Camilla's who started with us in St. Jean. I have not seen her since that first day, and the transformation is startling. My initial impression of Jean was an angry, unhappy, quiet woman who wanted nothing to do with any of us. This woman is glowing, talkative and has shed at least 20 lbs. She quickly separated from our gang in the beginning, but I will soon see that she has found her own Camino family.

I ask if I can join their group of five, gathered around the communal table outside our albergue. They all smile and nod as I sit and place my wine on the table. Jean sits opposite me, next to a cheerful, handsome, grey-haired man. She puts a possessive arm around him and introduces him as Paul, from England, her *Camino husband*. Paul laughs a little nervously and is quick to mention his "loovely" wife back home in England, in his thick Northern-British accent. He appears to be a little uncomfortable with Jean's assumption

The Camino wife and husband thing for me is like biting into a dessert in Japan that looks like it's filled with chocolate cream, only to find it's sweetened adzuki bean paste. I have to fight my gag reflex, resist the urge to spit it out, and instead swallow it and pretend it's delicious. This pilgrimage is such a unique opportunity to spend time with yourself to heal and grow, I can't imagine why people want to attach themselves to someone in this way. There is no doubt that deep, lifetime friendships and bonds are formed, but to liken them to a marriage? Still, I get the message, loud and clear, "*hands off.*"

It's relatively easy to find common ground with a group of strangers on the Camino. And so, we move past introductions to discuss how the energy for this last 100 kilometers is very different from the beginning. In part because of all the self-proclaimed, Camino Lites. I had encountered this phenomenon the day before, on the cow shit day. I watched a large tour bus drop off a group of German tourists, with no day packs, not even water. They strolled for a couple of kilometers, talking loudly and excitedly amongst themselves. They eventually stopped for lunch before being picked up by the bus again. In an effort to avoid judging, and not let them spoil my experience, I just tuned them out and let them get ahead of me.

Last night over our drinks with Gordon, Rene told me of running into June, one of our English friends, shortly after the bus drop off. Remember that she is the one who had to taxi through five days of her Camino due to severe blisters. With a lift of her nose, June in her *proper* English accent had said, "Did you see that busload of people?" – *sniff* – "makes you feel a little superior, doesn't it?"

"She actually said it." Rene had snorted to me as we both doubled over laughing.

As I finish my drink now, I excuse myself to find a quiet space to write in my blog. While they are all lovely people, I sense that, with the exception of Paul, who is exceedingly joyful and friendly, this is a *closed* group, not looking for any new members. That aside, they did invite me to join them for dinner that evening across the street and were all very kind and courteous.

Another difference at this stage of the Camino is that when starting in St. Jean, everyone is fresh, excited and looking forward to meeting lots of new people. As the journey continues, friendships form, and with tacit agreement, some groups *close*. The initial urge to build new acquaintances has been replaced with allowing the comfort and safety of the group or person, to hold space for deepening of the experience.

Day 36: Melide

It is such a privilege to be able to walk for so long, that snow turns to Spring, and that in turn becomes Summer. There is a definite difference in not only the temperature but also the feel of the air today. The morning mists of Galicia (pronounced Galithia by the locals) have a timeless, fairy-tale-like quality to them, blanketing everything in an otherworldly silence. I leave in one such mist this morning, at the same time as the group from last night. The women all greet me, then pair up and stride off. Paul notices that I can barely walk as I wait for the ibuprofen to kick in. Though he has enough energy for the two of us, fairly bouncing on the balls of his feet, he hangs back to keep an eye on me. I try to release him from service, I don't wish to separate him from his friends, but he quietly ignores my entreaty to go on ahead. He walks out in front for a while but keeps coming back to check on me. I reflect on how some men love a damsel in distress. I've been such a powerful, independent woman for the past two decades, it's no wonder I can't find a date. I can see that my new-found vulnerability has a power all its own.

The first tiny village is only two kilometers away. We stop at the only open bar for breakfast, to find the rest of Paul's group saving places for us at a long table. Paul orders toast with cheese and tomato. He knows no Spanish, and his British accent is so thick, even I have a little difficulty following him. Unable to understand him, the waitress with a shrug of incomprehension, plops a slice of Swiss cheese on top of the strawberry jam she has spread on his toast. Paul is too polite to return it (and I'm not sure that's even an option here), so we chalk it up to our first laugh for the day.

As we get ready to move on, I turn to see the Irish couple come down the stairs in the back of the room. They must have stayed here last night. They are so happy to see me and greet me exuberantly.

After we leave the bar, the group quickly moves out ahead of me, with the exception of Paul, who hangs back a little, to keep an eye on me I presume. As I walk down one magical corridor of trees, the morning light creates a golden glow over everything. Paul, 30ft ahead of me, waits for me to catch up with him.

"Do you notice the *loovely* light here this morning? Isn't it splendid?" He asks in his thick accent

"It takes my breath away," I say with moist eyes. "Let's take each-others picture, see if we can capture it." By some miracle, the picture did the light justice, it remains one of my favorite images. I feel a wave of gratitude roll over me for this entire experience, and that I have the resources (time, money and healthy body) to do this.

The rest of the group is out of sight on the wooded path. Paul moves on ahead, but thankfully is in shouting distance, when I notice a tall, young man with a brand-new backpack walking towards me. A red flag goes up, but I am comforted that Paul is near enough to call if there's trouble. Sure enough, as he passes me, in heavily-accented English, he asks, "Speak English?" I shake my head and quickly say, "I'm sorry, no." I immediately realize what I've done and chuckle under my breath – I would make a terrible spy, I crumble under pressure. Whatever – he gets the message and keeps walking.

I go around a corner in the woods and see Paul waiting for me at a large puddle of water covering the entire trail. Some flat, uneven, slippery rocks create a tenuous, dry path over it. He is sticking around to help me across safely, as he can tell I'm still not fully *in* my feet yet. I feel nothing but kindness, from this beautiful, chivalrous man. He is my self-appointed Camino Angel today.

We catch up with Paul's friends at the next coffee break. I sit with them for a while as we all enjoy our second breakfast, but their lack of desire to converse with me is just short of rudeness. Paul is a bright light, they don't want to lose him from their group, and I don't blame them. I sense they feel threatened. I take the hint and insist that he leave with them.

I have been walking by myself for some time when a bar materializes. I'm so ready for a break, and this place is a welcome beehive of activity. As I stand in line to get my new favorite food on the Camino, Gallego soup, I spot Missy talking with the group of young people who've adopted her. The smile she returns when I wave is half-hearted, she looks worried. After I order, I walk over to extend some friendship. She seems a little lost, my heart goes out to her. She appears to appreciate the effort, and says she will come talk to me once they get done exchanging contact information. The international group have their heads bent to their phones, as they try to figure out country codes.

I set off after lunch, the yellow arrows soon direct me off the narrow roadside trail onto a sandy, wide path leading up through farmer's fields. Missy catches up to me, and we make small talk. But inevitably her longer stride and our lack of genuine connection, have us amiably bidding each other adieu. I will not see her again until Santiago.

An old man bent over tending his garden attracts my attention, as in my peripheral vision I see an old woman walking towards me, carrying bags of groceries in both hands. Her sensible, low-heeled shoes remind me of my grandmother, with a calf-length dress and full apron tied around her waist filling out the ensemble. She is backlit by the sun, creating a glowing, ethereal vision. One of life's ordinary moments, elevated to the mystical. She looks at me from underneath her broad-brimmed hat, and we make eye contact. Neither of us looks away. We both smile and in unspoken agreement stop. With hand gestures and the few words of questionable Spanish I know, I try to express the beautiful vision she creates. It's ok that she cannot understand me, as there really are no words for what I want to convey. She, in turn, tries to tell me something in her native language. Her face, deeply lined from a lifetime of smiles, lights up and tells me all I need to know. Our mutual appreciation is evident. The language barrier prevents us from logically understanding each other, but this exchange is happening on many levels. Finally, reluctantly, she says, "Buen Camino." As I walk away, the word *Grace* comes to mind. I feel I have been kissed by an angel.

Today I walk endlessly through villages so small, sometimes I wonder if it's just someone's backyard. I have one more marathon day tomorrow (25.5 kilometers – God help me) then two short ones. Unfortunately, in Galicia there are some long stretches with no accommodations, so I have to work my itinerary around them. I'm frankly a little scared of this big day tomorrow, as my feet are miserable. But I'm hoping for a miracle – only 52 kilometers to go. I will crawl into Santiago if I have to.

Melide is a large, sprawling town, and there is a steady uphill climb to get to my hotel, which is on the far side. At first-glance, the run-down building I have a reservation at looks depressing, but I am taken around the back, and once inside it is quite nice. Up three flights of stairs, ugh! But the private bathroom with bathtub raises my spirits. After a long soak, all is well with the world again.

I receive a text from Rene, she is also in Melide and wants to meet for dinner. I'm happy she's forgotten about our *Camino reset*, so I don't mention it. We are at different ends of the town, about a kilometer apart, so we decide on a bar halfway between us.

The food specialty in this part of Spain is pulpo (squid). Rene appeals to my sense of adventure, and I agree to split an order with her. Chunks of off-putting squid legs, complete with suckers on the bottom, are slathered in olive oil and nothing else. I appreciate their beautiful presentation on a wooden platter, and tentatively spear one with my fork. They taste like greasy pencil erasers, and I'm having a hard time wrapping my mind around eating what I know to be a very intelligent creature. I'm far from being a Vegan-Nazi, but sometimes my love of animals interferes with my enjoyment of eating them, go figure! It's not the only reason I've been a vegetarian (occasional vegan) for the past two decades, but it's one of them. I try putting salt on the second chunk to see if that helps. Well, that one tasted like a *salty,* greasy pencil eraser, and this time my gag reflex rears its ugly head. That does it, they don't even taste good. I reach for a slice of the ever-present fresh bread from the full basket and decide that will be a perfect dinner after a long day of walking! Besides, with enough of this delicious, homemade Sangria, I'll stop caring pretty soon.

Rene and I laugh and share stories about our walk today. I think what I miss most about not having a husband or boyfriend, is having someone to witness my life. Downloading and discharging experiences at the end of a day to someone who cares, is a precious gift.

Two tables down, I notice the group from last night with Jean and Paul. I excuse myself to go say hello. Paul gives me a genuine smile, but the welcome from the rest of them is lukewarm. Sheesh, alright already, I get it. They've closed up ranks, not taking new members. I just wanted to say hi for goodness sake.

The walk back to my room for the night is long and lonely, in what feels like a sketchy part of town. I'm seeing that the trade-off for sleeping in a hotel instead of an albergue, is privacy for companionship. The camaraderie of being surrounded by people with a common goal, even if they do snore and fart, is comforting in a different way than a good mattress and a bath are.

Day 37: Salceda

I am relieved to put this hotel behind me and be back on the trail, with the promise of seeing other pilgrims again soon. It's 6:18 a.m., I want to get a jump on this long-ass day. I love to walk out of these big towns just before sunrise when the yellow street lights cast their magical glow.

The air is scented with eucalyptus, as the route takes me through one wooded area after another of tall, ancient trees. I will spend a lot of time completely alone in the forest today, with very few pilgrim sightings. After just half an hour in the woods, a peace reminiscent of watching a newborn sleep, permeates my body. A stop on a bridge over a gurgling creek, makes me wish I had time to meditate. But this is a long day, and I need to keep walking.

I'm drawn into a timeless zone on this intoxicating aromatherapy stroll. In the small village of Boente, an enterprising person has positioned their bar as the first place to greet you. A waitress brushes the leaves off an outside table and straightens the chairs, while balancing an empty tray in the other hand. I need no further invitation and unclip my backpack. I set it down with my poles at the table she has just cleared with a, "Gracias." Even though there is a slight chill in the air, I can't bring myself to eat inside this morning.

The smell of bacon greets me as I push open the door to go in to order. I watch a plate of it with two eggs pass me and decide I will have that as a change of pace from my usual tortilla. But I have to wait in line for five minutes – long enough to reconsider. I haven't eaten bacon in 20 years – hmmm, I'm not quite ready to go there yet. Eggs and toast it is.

I am re-seated, listening to a cuckoo sound off the hour when I see Rene approach. She has a big smile on her face, her hiking skirt swishing from side-to-side as she strides confidently towards me. I take a sip of fresh-squeezed orange juice and close my eyes for a moment, as I relish the cold, sweet deliciousness. Every cell in my body feels more alive today, everything sounds, smells and tastes better. There is a cumulative effect to this magical journey.

Rene puts her backpack on the chair next to mine, admires my breakfast, and goes in to order one just like it. Our spirits are high, as we can feel the end in sight. We have discussed our mixed feelings, but for both of us, the thrill of impending accomplishment predominates. We leave together after a leisurely meal and walk for ten minutes down a steep street. We are behind some young, female pilgrims oohing and aahing over a sweet little puppy, who now wants to follow them. *Good luck with that,* I think to myself and make sure to give the puppy a wide berth.

I can feel Rene itching to walk faster, as we continue to descend. I release her with a smile and a mock military salute, stiff hand to eye, "I'll see you at the next café," I say. She returns the salute, and with an enviable spring in her step, I am soon eating her dust.

As I leave the village, a Spanish man, with a full head of dark hair streaked liberally with grey, walks uphill towards me with a fully bloomed, red rose in his hand. He passes a young, pretty girl in front of me, nods at her and keeps walking. As he approaches me, an ear-to-ear grin lights up his face, and with a flourish and a small bow he wordlessly presents me with the rose. I later find out, he had first passed Rene who had hoped to be the recipient of the beautiful flower. But he waved it at her with a smile and an, *I'm-sorry* shake of his head. I'm suddenly lit with joy. Can I see myself as that rose in full bloom, still beautiful and desirable? I tuck it in the shoulder-strap of my backpack so that I can lean over periodically to smell it.

I feel rejuvenated as I cross the street to enter the next part of the forest. My feet are holding up so far, and I decide for only the third time since starting the Camino, I would like to listen to some music. I have a really long haul today, and I hope it will help the time pass. But what do I want to hear? I get an urge to put on Pandora and look for 80's love songs, a little voice in my head says, *nobody does love songs like the 80's.* I shake my head, where did that come from? I've never listened to that on Pandora, in fact I only listen to Pandora when I'm doing a massage. I don't question my intuition though, I look forward to a little trip down memory lane.

Completely by myself in the forest, and five songs in, I bend my head to smell the rose and stop dead in my tracks. My eyes fly wide open, my jaw literally drops! Time stands still as I'm flooded with a huge realization. Oh. My. God! This journey has been about healing my heart. A big smile spreads across my face, and I laugh out loud to the trees. I am so dense. How could I have freaking missed it? The message is so strong, if I didn't have so much ground to cover today, I would sit down and be with it for a few minutes. But instead, I allow integration to happen while I continue walking.

Rene and I have talked about how much male attention I have attracted on this trip. I certainly did not come looking for it, but they've been showing up since the first day. All very positive experiences, with kind, gentle, respectful, caring men. On reflection, I see that some were just friendly, but others definitely had a more romantic energy. As I continue to walk and think about them all, I can see that several represent different men I've been in a relationship with over the years. The relationships didn't end so well, but the brief interludes on the Camino have all completed on a positive, uplifting note. It's as if I've re-written history.

My last boyfriend was 13 years ago, with not even a date since then, so you can appreciate my surprise at this strange turn of events. With the luxury of time on my hands, as I step carefully along the forest path, I go over in my head some of the other facets of this journey: Long walking meditations every day. I've allowed all of my eating and other habits to relax and change. I sleep in a different bed every night. I feel vulnerable all the time.

I look around and notice how completely alone I am, except for my friends, the trees. It feels like a complete shakedown. The walls around my heart built from years of failed relationships, could not withstand the vulnerability caused by the loss of my daily routines, designed to keep me *safe*. Combined with the one-two punch of so much male kindness, they've crumbled, and my heart has softened and healed. How do I know this? I just do. I can feel a subtle but very real difference in my feelings towards men right now. A trusting, and openness to possibility that has not been there in a very long time.

No more than five minutes later, as I continue to walk and integrate my *Come-To-Jesus* moment, I look down in front of me, to see that someone has fashioned a large heart from sticks, then filled it in with fresh, green moss. It's 18" across, smack dab in the middle of the path, hard to miss. Talk about Camino magic! My angels are not messing around today.

I notice that the concrete markers that are counting down the kilometers to Santiago have all been vandalized, the remaining kilometers removed. So, with no clue as to my progress, I continue to walk, trust and surrender. A thought pops into my head; I wonder if all the pain I have been experiencing in my feet, is a physical manifestation of the emotional pain I have been carrying around in my heart, unawares, for so long. A lifetime of love and loss. If so, I ask my angels to please release it.

Like the tortoise and the hare, in spite of our different paces, Rene and I never seem to be far apart. Miraculously, we meet again at lunchtime, and I share my magical morning with her. It's late afternoon, and we don't want to lose our places at the albergue we have booked for tonight. So, Rene leaves ahead of me, her short, sassy hiking skirt swaying atop her rainbow socks, a vision of *joy* as she walks into the forest. She will let them know I am on my way.

As I pass through the next village, I notice an ancient stone house, with patches of concrete repair, the shutters painted a cheerful kelly-green are thrown wide open, dishes sit on the windowsill. I suppose if there is still life in these old buildings, there could still be time for me to confront my Final Frontier – relationships.

On top of this being a marathon walking day, the private albergue we're staying in tonight, Toristico de Salceda, is a little difficult to find. *"You need to make a left at the intersection in the woods, with the arrow I made from sticks on the ground."* Rene texts.

I chuckle/snort/sigh I'm too tired for Hawaii-style directions. No one uses street names and numbers back home. You have to look for bamboo patches, questionable gates and the third driveway on the left – wait, is that a driveway, or a farm road for someone's coffee field? – does that coffee shack count as a house?

After passing several intersections in the forest with dubious sticks in the dirt, that could have once been an arrow before someone potentially kicked them, I start to wonder if I've gone too far. How will I know? My feet are hanging on by a thread.

Praise the Lord! That's definitely an arrow in the middle of the path. Thank you, thank you, thank you! My *Hallelujah moment* is a little premature, however. The albergue is still almost, a kilometer away.

I started walking at 6:18 this morning and arrive at this oasis at 4 p.m., with only a few short breaks. As a reward, Rene and I have lucked out tonight. This place is new, modern, and has a patina of luxury, complete with a sizeable foot-soaking pool. I can't get my sandals off fast enough.

The pilgrim's meal is served in an elegant, glass-enclosed dining room, costing no more than meals served in more humble surroundings. I am so grateful to the Spanish for their deep respect honoring those making this pilgrimage, and their ensuing service to us. I don't think I've ever seen anyone work as hard as the hospitaleros at the albergues across Northern Spain. The owners often attend to every aspect of the business from the paperwork, to housekeeping, to meals. They are up early to make breakfast and to bed late after clearing up dinner, seven days a week. With only a few months off in the winter.

I feel immense relief at being only two short walking days away from Santiago. I really had my doubts that I would physically be able to walk 25.5 kilometers today. I tend to my feet with the usual creams, self-massage, and compression stockings. But before I go to bed, I pray that if this pain is in part, a reflection of the emotional pain in my heart that I recognized as being healed today, please let it be relieved tomorrow. It went something like, "...can you please cut me a break, God?"

Day 38: Villamajor

Breakfast is served downstairs, and by Camino standards is not bad, meaning: at least there is one. I leave the box of orange juice for someone else, I have become far too spoiled by the fresh juice available in all the bars. The croissant in a plastic sleeve looks edible though, I ply it with strawberry jam from the minuscule, peel-back container. I silence the *conservationist* in my head that cringes at all the unnecessary plastic wrapping. But most importantly, no self-respecting Spaniard would consider serving café con leche any other way than hot and fresh. So, the carbs are consumed for fuel, the coffee for enjoyment.

"How are your feet this morning?" I can see Rene is almost afraid to ask. On her last Camino, she was cursed with painful blisters, this time she is blessed with very few physical ailments, walking virtually pain-free.

"Shh, let's not jinx it," I say with a smile, giving her the answer.

Our busy host does double duty, serving breakfast and checking out pilgrims, eager to be on their way. But he takes the time to give us hurried directions on the back of a torn off piece of paper for a *shortcut* to return to the Camino. Directions wouldn't be complete without him abandoning his register, walking outside, and pointing to a path in his backyard leading away from his property. He waves his hands and speaks in Martian or Spanish, I'm not sure which.

When you each only speak a little of the other's language and rely on copious hand signals and a quickly scribbled map for directions, there's fertile ground for misunderstanding. We are soon hopelessly lost and wandering aimlessly on country roads. After we try several lanes that turn out to be dead ends, we stand in the middle of a deserted road. Rene swivels around 360 degrees consulting the scrap of paper. I have my head bent over my phone trying in vain to pull up apps.

"Damn, there's no cell coverage here, I have three map apps on my phone, none of them will download, we're going to have to go – *dong, dong, dooooong* – old school." We both chuckle and the slight tension is diffused. I put my phone back in my hip bag, Rene stuffs the *map* in her pocket, and we start to walk in what we think is the right general direction. We are relieved ten minutes later when we see two pilgrims cross a road ahead of us, shortly followed by a yellow arrow sighting.

I am in very little pain today and could keep up with Rene if I wanted to. But with only two days left of this precious journey, by consensus, we both agreed earlier that we would like to walk alone. Now that we are safely on the right path, Rene says, "I'll see ya later," and with long strides, she's soon bouncing out of sight.

I can't help but wonder what this trip would have been like if I could have walked normally and not been in daily pain for three-and-a-half weeks. I see that it would have been completely different. I would have met different people, had different experiences – I quickly dismiss the conjecture as fruitless, and am instead grateful for each minute of this journey. Knowing that in some strange way the pain was a part of it.

I am prepared for the walk to become less rural as we get closer to Santiago and the airport. But the Spanish have done an amazing job of preserving the integrity of the original path. They have left just enough natural greenery around it to create an illusion of still being in nature. We walk through many patches of eucalyptus forest again today. Yes, *we* – although Rene and I want to walk alone, we keep running into each other at coffee and lunch spots. Resistance is futile, so we just give the other plenty of space and don't talk too much.

Several days ago, Rene and I had a discussion about her female urination device (a FUD – no comment). It's shaped like a little funnel and allows a woman to pee standing up. I had heard of them and briefly considered getting one, but figured it was one less thing to carry since I have no problem squatting. We find ourselves momentarily alone on the trail beside a large tree.

"Oh, good I need to go pee," I say.

"Me too, let me just whip out my dick," says Rene – I lose it! I laugh so hard it comes in waves. Now I'm laughing just because it feels so damn good, like some cathartic overload of joy. Tears are streaming down my face, I can't stop. Laughter being contagious, Rene catches it, and the two of us laugh and cry together. I am barely able to keep my balance as I squat on the balls of my feet, now wishing I too had a *FUD*. As I pull up my pants, I look at her with a straight face, she looks at me, we lose it again. I hold my stomach, which aches from laughing. Damn, I love a good laugh!

We both agreed we wanted to take our time today, clean our plates of every morsel of this delicious experience, quite literally. So, we stop at most of the villages we pass through. Ensalada mixta (mixed salad), my go-to lunch, tastes especially good, as we join others at a large outdoor seating area. I watch a steady stream of pilgrims walk up the trail, some stop as we did, some keep going disappearing into the tall, majestic eucalyptus trees.

We enter the forest of aromatherapy giants once again. We come to a fork in the woods, and an enormous, magnificent Grandmother-tree is glowing as the sun hits it just right. Its presence feels ancient and magical.

Our albergue in Villamajor is further than we thought, as we walk through one village after another that is NOT it. The good news is, it will be a shorter distance on our last day tomorrow.

Because we have more time today and we're trying to prolong the inevitable, we make one more stop. As I wait to order a drink, I see a beautiful bocadillo being handed across the bar. I discover it is thick slabs of local bacon, tomato, and lettuce, my mouth waters looking at it. I haven't eaten pork in over 25 years, neither has Rene. I hesitate, but with only two days left of my *beliefs/routines/habits vacation*, I decide to go for it and order one.

"Your piggy sandwich looks good," Rene says, as I sit down.

"Why don't you eat pork?" I say. "I notice you're not a vegetarian."

"After my parents sent-to-market, that's how they sugar coated it, *sent-to-market*, my prized 4H pig that I had loved and nurtured for years, I couldn't stand the smell or thought of eating pig again."

I wave the sandwich under her nose, "oh, but doesn't it smell good now? Come, join me on the dark side," I entice her.

"Ok," she laughs, "just a small piece." We push our guilt aside and thoroughly enjoy this carnivorous treat and honor the pig with gratitude. I feel somehow liberated.

Half an hour later, leaving yet another small village, assured that the next is ours, we are confronted by a monstrously steep and long hill, snaking endlessly up for two kilometers. The cyclists that pass in first-gear are clearly struggling too and not moving much faster than we are.

"You have got to be freaking kidding me!" I say, laughing out loud at the relentless climb ahead. I look back to commiserate, but Rene is lagging behind on a phone call.

As I slowly plod up the hill with no breaks, something I could not have done a few weeks ago, I am in awe at the newfound strength of my body and mind. I also have speed and rhythm down to an intuitive science.

We eventually arrive at the private home we booked for tonight. Our room turns out to be a tiny apartment with a small sitting area and full bathroom. When we checked in, the young man slid a tiny slip of paper across the counter with the WiFi password on it – that was wishful thinking. I have walked every inch of the room and property with my cell phone held out in front of me and watched it flicker from *No Service*, to one bar. I smile, it makes perfect sense that this evening should be quiet and reflective. The energy is a little heavier than usual, as we both contemplate the reality of this journey ending. It will be bittersweet for sure.

Day 39: Santiago de Compostella

We are up early this morning, eager to finish our adventure. My feet and legs feel really good. But I don't take the time to dwell on this miracle.

Rene and I decided a few days ago that we want to walk into Santiago together. It's a special moment. Each person has their own reaction. Some even report it is anti-climactic after all that you go through to get there. So, I try not to have too many expectations. But I do know I want a friend to witness it with, and Rene has become a very good friend in a short time.

The first couple of kilometers are uninteresting as we walk through quiet neighborhoods, until we reach the famous Monte do Gozo. A beautiful sculpture adorns the site where we are supposed to be able to get our *Hallelujah Moment* of the Cathedral, hence the name Hill of Joy. But I am unable to sing out my last refrain, as trees and buildings now obstruct this, once beautiful, view and there is little else of interest here. The food truck parked next to a one-room chapel sells styrofoam-cup coffee. I purchase and discard it after one sip. I take a second look at the chapel and realize that while unremarkable, it feels authentic, so I stand in the tiny room by myself to say my last prayers for Cooper, on the trail.

We are now very close to the City, so we are soon in heavy traffic. We cross a bridge and carefully navigate a round-a-bout. The bar we stop at for breakfast on the outskirts of Santiago, is modern. It has nice wooden tables that don't wobble, set up neatly in a linear fashion, and waitresses that come to your table. It's so civilized! I already miss the quaint, unique bars scattered across the countryside, all with their own personality. Gone is the bustle of excitedly, chatting pilgrims lining up at the counter to order their food and drinks, backpacks and hiking poles strewn everywhere. We sit silently, each contemplating the end of this magnificent adventure.

I look through the window, and walking past is my German friend Gudrun, who I haven't seen in two weeks. She looks in as she is passing and we both scream soundlessly through the glass. She comes into the bar and we girly-hug excitedly like old friends. I remember now that she was the woman sitting with Rene on the blanket the first day we met. What a perfect, full-cycle completion.

Rene starts to get up to leave, "I'll see you in Santiago," she says with a *not-quite* smile.

I'm momentarily puzzled, but then realize she has misread my excitement, and in the fragility of our new friendship is trying to be considerate. While I am excited and surprised to see Gudren, Rene is the one I want to experience entering Santiago with. "No, no, hang on," I hold my hand out to stop her. There's really not much to say after the exuberance of our greeting, so Gudren says goodbye for the last time, we will not see each other again.

Café con leche has become a time-honored ritual, that unites pilgrims from all around the world on the Camino. Quietly we pay the check, aware that this will be the last of this magic elixir on the trail and rejoin the busy sidewalk. We come upon a pilgrim named Rose that Rene knows, and invite her to walk in with us. As we begin to traverse the narrow, crowded sidewalks in the old part of the City, I look up and sing, "Hallelujah," for the last time. Rose looks at me quizzically, Rene smiles and follows my gaze. The tower of the famous Cathedral, our destination today, is finally visible peeking above the rooftops.

We all stop. My heart is pounding, and I begin to softly cry, which sets Rose off too. Rene is misty-eyed as she re-lives the last time she did this, she tells us it's a different experience each time. After a teary minute, with a mischievous smile, she points across the street and says:

"If you want to break up this touching moment, check that out..."

My eyes follow her finger to a white, miniature poodle, stopped in the middle of a busy, narrow sidewalk. He assumes the classic squat, as he unashamedly does his business while people step around him. His owner on the other end of the leash looks everywhere but there. Well, that was a buzz kill, now we're howling, with tears of laughter streaming down our cheeks.

As we get closer to the Cathedral, we lose sight of it and just keep following the crowd. Before us is a tunnel with a man playing bagpipes. We exit the tunnel into a huge square and to our left, the Cathedral looms majestically above us, shrouded in blue scaffolding. We laugh and cry, as we take the obligatory pictures.

I see a familiar face and recognize the group from Australia who I have been leap-frogging since the first day at Orisson. I walk over to hug and congratulate them.

I hear my name called and turn around to see Alaia, a friend from the Big Island of Hawaii who I did some training with at home. She started before me and ended a week ago. What are the odds we would meet here? Our greeting is surreal. Alaia would pass away 18 months later from ovarian cancer. Although she had never mentioned it to me, I suspect she knew about her illness then, and this journey was a magnum opus to her life.

Standing next to Rene again, I look around me 360 degrees. "So, where's the Parador?" I ask. I'm excited to see this legendary, five-star hotel. We decided several days ago that we would splurge for two nights when we arrived in Santiago.

She points to the left of the Cathedral at an imposing, rectangular, gothic building. Now I notice the word *Hotel* in gold letters at the top of the, almost ordinary looking, glass entry doors. Perfect, understated elegance. Like *old money*, this building doesn't need to prove anything to anyone. It turns out that its' exterior is as humble, as its interior is palatial.

We take in the grandeur of the vaulted ceilings of the lobby and the deep, richly polished furniture, with brocade upholstery – I'm guessing there will be no wobbly tables here, as we claim our reservation. Time appears to have slowed down, possibly as a result of the unhurried progression of the past 39 days.

This place is enormous, with a couple of courtyards and at least two buildings. When we finally find the room, Rene does some google digging and finds out that the Paradors across Spain are historic structures that have been restored by the government. They are then run as five-star hotels to pay for the continuous upkeep that ancient buildings require. It's also clear to see that they use these places to house antiquities, it's like living in a museum.

I get a text from Kevin,

On my way to Finisterre, planning to bus back tomorrow. Let's meet for dinner?

Sounds good. Rene and I staying at the Parador, I'll have to show you our digs.

Finisterre (End of the World) is an additional 90 kilometers past Santiago. While many choose to make this optional journey that goes to the ocean, I'm just grateful to have made it this far.

The culmination of this pilgrimage for most is the attending of a Pilgrim mass, held twice daily, in the famous Santiago Cathedral. We arrive early, so I leave Rene sitting in a pew and walk around to admire this magnificent, architectural feat. I run into Deidre and Camilla, two of the four women that I started with 39 days ago. We hug, jumping up and down, and laugh. I ask them if they have walked through the Holy Door yet, which Rene and I had already scoped out. They look puzzled, they've been here a day and didn't know about it. So, eagerly, we all go out into the sun, around to the side of the Cathedral, and walk through together. What a wonderful, full-circle completion with these friends that I have kept in touch with daily via WhatsApp. We are all a little the worse for wear, tired and sporting fewer pounds on our bodies. But our joy at this shared experience has no words.

In spite of traveling on my own, I realize how many people I have befriended on this trip. Some I met at albergues, some I walked with for a while, many I then kept the connection with and exchanged messages daily on my phone. I recognize how important a feeling of community is for me and I am so grateful I was able to create my own version of a Camino Family.

I do feel a momentary sadness that Mary Jo and I had an untimely end to our budding friendship. It was so promising. I was surprised I didn't run into her and David in Santiago. I wonder briefly if I could have done anything differently. But deep down, I know that our *break-up* was a joint decision, and that the initial drama was crucial to triggering the chain of events that lead to my eventual heart healing. This is what I believe to be the real, albeit unknown, purpose for this trip. So, at the end of the day, I can thank her for being willing to assume that role in my play.

When I get back to my pew, Rene points out two of the English group, June and Kate-of-the-dirty-socks, a few rows back. I look and wave. Rene leans over and half-whispers, "Victoria is sick in her room, and Colin and Sophie are walking to Finisterre."

"And look who's sitting right in front of them," she adds with an almost imperceptible, side nod of her head, and raised eyebrows. I turn and look square in the eyes of Missy, who sits next to a handsome, blonde, European-looking, young man. She gives me a sheepish, half smile.

I'm a little disappointed that they don't swing the Botafumeiro today, a huge brass frankincense burner. Several times a week it is swung from one side of the Cathedral to the other by half-a-dozen monks pulling on ropes, three inches in diameter, that descend from the ceiling. It was something I was looking forward to, but to be honest, the enormity of this trip has put many things in perspective.

Back in our room, I finally get to wash off the day's journey. The famous Parador does not disappoint. The bath is sheer heaven, I lay there until the water cools. For the last time, I rinse out underwear and socks in the bathroom sink.

We go to a sweet little seafood bar and get an overpriced, but delicious NON-pilgrim meal. Then we get lost in our hotel trying to find our way back to our room. We laugh at the irony that we just navigated our way across an entire country relying mostly on hand painted yellow arrows yet keep losing our way in our hotel. Where are those yellow arrows when you need them?

Christine and I in Santiago

Day 40: A day in Santiago

Today is a *girlfriend* day of shopping and pampering. Rene and I are both tired of wearing the same clothes for 39 days. On a whim, we decide to indulge our poor feet and get pedicures as well. We are directed to a salon that can take us at the same time, and somehow, end up getting *the works*. After a mani/pedi, we are talked into a deep conditioning hair treatment. I feel the attendant smoothing something on my face, as I lay with my head tipped back in the sink.

Like a ventriloquist, with eyes closed and lips barely moving, I say to Rene, "Are they putting makeup on us?" The women working on us don't speak English. Rene laughs, "Affirmative! What did we agree to?"

After a hair blow out, and bright red lipstick application, we look in the mirror at the transformation. I guess you could say we look beautiful again, but I've become rather fond of my au natural look and have to wipe off the clown lipstick as soon as we leave. I'm not quite ready yet to assimilate back into this other reality.

It feels like the last of the Camino has been washed off us. I'm not sure how I feel about that. I've really enjoyed not worrying about how I look, hair pulled back tightly in a ponytail, wearing sort-of-clean clothes. This appears to be a sign that re-entry to the alternate universe that represents my old life, is imminent.

We stumble across a great little boutique were we both find a new outfit. There's a timeless satisfaction that comes from shopping with a girlfriend. Rene goes back to the hotel room while I continue walking around. As I pass a bar, my peripheral vision picks up some wildly waving hands. I look through the glass to see the faces of my Irish friends, that I met at Gordons a few days ago, beaming at me. They wave me in. I trip over a small step at the entry – ever graceful – and fall into the arms of a waiter, who winks and smiles as he catches me. Spanish men are so divinely sexy. My feet are much better, but I'm still a little unsteady on them (that's my story, and I'm sticking to it). I sit for a glass of wine, and we catch up on the past couple of days. I'm a little sad that I will never see them again.

Back in the room, Kevin texts me, *It took longer than I thought to get a bus back, won't be there in time for dinner. Wanna have breakfast tomorrow morning?*

Oh, good I can take my new outfit for a spin. *Absolutely, you have to come to the Parador, they have an incredible breakfast buffet.* I know this 6-foot 2-inch tall man can eat.

The next morning, I approach him in the lobby. My hair still looks good from yesterday's extravaganza, and I feel a little sassy in my new, different clothes. "Well, you clean up good," he says with a tired smile. I realize then, that while I've had a day-and-a-half to rest and unwind, he just finished walking. He has a little sticker shock at the price of the breakfast. For me it's included in the cost of my room, for him it would have bought two nights at an albergue.

After we eat, I take him on a tour of the hotel, including the five-star restaurant in the basement that is only open for dinner. We discovered last night on the internet that it was a former morgue for pilgrims who died here when it was a hospital. Rene and I amused each other with corny jokes, "I hear people are *dying* to get in," "I hear the desserts are *killer.*"

Both Kevin and I are headed to the airport this morning, but his plane leaves a little later than mine. I try to talk him into sharing my taxi with me, but he's planning on taking a bus, his frugality winning out. So, we say our goodbyes, promising to stay in touch.

Rene and I check out of the hotel. She is staying on another day and has found a cheaper place, this was a well-deserved splurge for both of us. We hug and say *A hui hou* (a Hawaiian farewell meaning – *until we meet again*). While some friendships you know are destined to end with the Camino, I feel I have made a lasting friend in Rene. Her sense of humor and down-to-earth honesty made her a perfect traveling companion. I will miss her. She has offered to help me with the book I have decided to write. I assure her that I will take her up on that.

I'm moments away from boarding my plane when I hear, "Angela!" I look up to see Kevin's warm smile coming at me. Too funny, the Camino is not done with us yet. We hug one last time, promising to stay in touch as he heads back to California for now.

Epilogue

The sense of deep peace I acquired on the Camino has, for the most part, stayed with me. I still think of it daily, and it often visits me in my dreams. I took off an additional week when I got home for integration and re-entry. But still, on my first day back to work, a student had a question, a client wanted to book another massage and the phone rang, all at the same time. I found myself putting my hands up, sitting back calmly and laughing. The client and student looked at me, then each other, silently. Their raised eyebrows and puzzled frowns told me they wondered what they'd missed. I could no longer multitask. My brain felt as if it had been re-wired, and I liked it that way. I decided to embrace this present from the Camino and not strive to return to my previous coping methods.

Perhaps the greatest gift of these 40 days however, was the way my life had been reduced to such stupendous simplicity. On the walk, the only decision to be made in advance each day (which took 10 minutes) was where to sleep the next night. Once the reservation was made, I put it out of my head. All other decisions regarding eating and resting were made in the moment and mostly out of my control. They depended on the location of villages and cafes.

The deep desire for simplicity and peace that our souls long for in the hectic world we live in, is I think, the common denominator attracting many people to the Camino today. The challenge of course, then becomes how to hold on to that when you return home. Over the course of the next year, I decided that the main source of stress in my life was my school. So, changes happened. I hired more people, delegated more, and spent my extra time at home writing this book. Through this process I gleaned even more lessons from the journey.

To answer my most frequently asked question, Kevin and I talked on the phone a few times the month after I returned. He's starting a new life in California, so it would appear that the potential of anything more than friendship was slated to end with the Camino. But the shared experience of this monumental walk is hard to completely let go of. We continue to keep in touch with texts and emails whenever we think of each other. I am eternally grateful to him for the part he played in reawakening pieces of me I thought were gone forever.

Rene (not her real name) and I have continued to deepen our friendship, and she followed through on her promise to help me write this book. She walked her third Camino in 2017 with her husband, and we walked part of Camino del Norte together in the Spring of 2018.

Cooper finished his treatment protocol for leukemia in October 2017 and is in full remission. In five years, we can call it a cure. His artwork is on the back cover of the book, per my request to, "draw Nana some poppies."

I must admit, I did wonder if I might end up in a relationship again after all these years. But then I had the realization, that the healing of a broken heart from lost love, was not necessarily so that I would have another partner. It was more so that my heart could be whole again, not tainted by experiences from the past. My life is full of love and excitement, it's hard to imagine that anything is missing. I guess only time will tell. I have noticed though, that my general attitude towards men has softened and is more accepting, and I have a newfound interest in the possibility of something more.

But is my heart really whole again? As I fear, like many before me, I may have left a piece of it on the Camino.

Gracias

To my three daughters: Michelle, Alicia and Chelsea, I wish to say a big thank you and mwah for your unconditional love and support during my journey. Never once did you doubt I could and would do this. When I called you from the trail having a hard day, your encouragement would give me renewed strength. Thanks to the three of you for always being my true heart compass.

A big shout out to my friend and fellow pilgrim Christine. Beginning with a fated meeting on the trail, our friendship has expanded on many levels. We laughed so hard together, a welcome respite on some grueling days. Thank you for bearing with me in the writing of this book, I could not have done it without you. I started with what I thought was a great story to tell and a lot of enthusiasm. Your professional experience and wisdom helped me to channel it into a coherent book. With so much patience, you gave me a crash-course in Creative Writing, guiding me through the process of editing, re-writing and editing some more. My heart is eternally grateful.

To my friend Lynn, who has patiently listened to me first rattle on about taking this journey, then obsess with writing about it. Thank you for always picking up the phone and being that voice of reason for me, and for talking me down off the ledge when I went into fear and doubt.

Thank you to Douglas, for your friendship and your heartship. I dragged you along on countless training hikes, lured by the promise of breakfast. We have had so many shared adventures together in this life, it seemed fitting that you would in your own way, join me on this one. And thank you for your beautiful cover design for this book.

To my friend Dawn, who held down the fort at home. Thank you for creating a safe, loving space for my dogs and for holding me in your prayers.

And Deenya, thank you for always supporting my travels. Your behind-the-scenes work and willingness to jump in when needed doesn't go unnoticed and is so appreciated.

I would like to acknowledge the women I started this journey with. Deirdre, Pamela and Camilla (not their real names), I appreciate your continued friendship. A thank you to Deirdre for being one of my beta readers in the early stages, when this book wasn't even close to ready.

Thanks to Bill and Patty, two fellow pilgrims from Hawaii who share my love of the Camino. Our training hikes together helped to keep me inspired. And thanks to you two also, for being beta readers long before this book was polished.

And finally, my sincere, heartfelt appreciation to Kevin. Hopefully after reading this book, you might begin to understand the pivotal impact you had on my journey and my healing. I treasure our friendship and hope we continue our periodic conversations.

About the Author

Born in England, Angela's family emigrated to the US when she was 13, at the height of Beatle mania. She has three grown children and six grandchildren. Her youngest daughter accompanied her when she moved from Rochester, NY to the Big Island of Hawaii in 1994. She founded the Aloha Massage Academy in 2006, where she continues to teach massage. Angela lives off the grid with her two dogs, MacKenzie and Nui, in a *Tiny Home* in a secluded location up mauka (on the mountain).

The Holy Door in Santiago

Made in the USA
Columbia, SC
26 May 2022

60931548R10134